Topics in Computational Algebra

Edited by
G. M. PIACENTINI CATTANEO
and
E. STRICKLAND
Department of Mathematics,
University of Rome II 'Tor Vergata', Italy

Reprinted from Acta Applicandae Mathematicae,
Volume 21, Nos. 1 & 2 (1990)

KLUWER ACADEMIC PUBLISHERS
DORDRECHT / BOSTON / LONDON

Library of Congress Cataloging in Publication Data

Topics in computational algebra / edited by G.M. Piacentini
Cattaneo, E. Strickland
p. cm.
"Papers presented at the Computational Algebra Seminar held at the
University of Rome 'Tor Vergata,' 9-11 May 1990"--P.
"Acta applicandae mathematicae, volume 21, nos. 1 and 2,
October/November 1990"--P.
ISBN 0-7923-1117-5 (alk. paper)
1. Algebra--Data processing--Congresses. I. Piacentini Cattaneo, G. M.
II. Strickland, E. (Elisabetta), 1948- . III. Computational
Algebra Seminar (1990 : University of Rome "Tor Vergata")
QA155.7.E4T67 1990
512'.0028--dc20 90-27032

ISBN 0-7923-1117-5

Published by Kluwer Academic Publishers,
P.O. Box 17, 3300 AA Dordrecht, The Netherlands.

Kluwer Academic Publishers incorporates
the publishing programmes of
D. Reidel, Martinus Nijhoff, Dr W. Junk and MTP Press.

Sold and distributed in the U.S.A. and Canada
by Kluwer Academic Publishers,
101 Philip Drive, Norwell, MA 02061, U.S.A.

In all other countries, sold and distributed
by Kluwer Academic Publishers Group,
P.O. Box 322, 3300 AH Dordrecht, The Netherlands.

Printed on acid-free paper

Printed in the Netherlands

Contents

Acta Applicandae Mathematicae **21**: 1, 1990.

Foreword

This volume collects the papers presented by 10 invited speakers at the Semester of Computational Algebra which was held at the Department of Mathematics of the Second University of Rome 'Tor Vergata' and was organized by Profs. G.M. Piacentini Cattaneo and E. Strickland during the period January–May 1990.

The aim of the Semester was to give an update on interesting techniques based on algorithms in different branches of algebra, with the idea of emphasizing the computational aspects of each branch. For this reason, the speakers were chosen from among the foremost experts in the various fields. The general fields covered are group theory, commutative algebra, algebraic geometry, representation theory, ring theory, invariant theory, and supersymmetric algebra.

The editors believe that the Semester largely succeeded in reaching its intended objectives and, therefore, wish to thank the speakers and all those who contributed to this success. They want to express their recognition to the institutions who have contributed with their financial support: Ministero dell'Università e della Ricerca Scientifica e Tecnologica, Department of Mathematics of the Second University of Rome, Centro Matematico V. Volterra, and the Second University of Rome 'Tor Vergata'. Also, a special thank you to the Gruppo Nazionale 'Teoria dei Gruppi e Algebra non commutativa' of the MURST and to the Progetto Strategico 'Algoritmi connessi allo studio di strutture algebriche' of the Consiglio Nazionale delle Ricerche for their generous contributions.

Finally, we wish to record a special thank you to Prof. L. Accardi, Director of the Centro Matematico V. Volterra, for his cooperation during the various stages of the organization of the Semester.

The Editors:
G.M. PIACENTINI CATTANEO
E. STRICKLAND

Acta Applicandae Mathematicae **21**: 3–39, 1990.

Branching functions for winding subalgebras and tensor products

V.G. KAC* AND M. WAKIMOTO**
*Department of Math., MIT, Cambridge, MA 02139, USA
**Department of Math., Mie University, Tsu 514, Japan

§0. Introduction.

0.1. One of the basic problems of representation theory is to find a decomposition of an irreducible representation of a group with respect to a subgroup. Namely, suppose that we have a representation π of a group G in a vector space V and suppose that with respect to a subgroup S this representation decomposes into a direct sum of irreducible representations:

$$\pi = \oplus_i \pi_i, \quad V = \oplus_i V_i$$

Given an irreducible representation σ of S one denotes by $[\pi : \sigma]$ the number of representations of S equivalent to σ in this decomposition, and calls this number a *branching coefficient.* An important problem is to compute the branching coefficients.

A special case of this problem is the decomposition of a tensor product. In this case $G = S \times S$, S is the diagonal subgroup of G, $V = V' \otimes V''$, where (V', π') and (V'', π'') are some irreducible representations of S, and the problem is to compute the numbers $[\pi' \otimes \pi'' : \sigma]$.

0.2. In the present paper we study branching coefficients for positive energy representations of affine algebras. Let us recall the basic definitions in the "non-twisted" case $(r = 1)$. See [8] for details.

Let $\bar{\mathfrak{g}}$ be a complex simple finite-dimensional Lie algebra of rank ℓ, and let $\phi(.,.)$ be its Killing form. Fix a triangular decomposition $\bar{\mathfrak{g}} = \bar{\mathfrak{n}}_- + \bar{\mathfrak{h}} + \bar{\mathfrak{n}}_+$, where $\bar{\mathfrak{h}}$ is a Cartan subalgebra and $\bar{\mathfrak{n}}_\pm$ are maximal nilpotent subalgebras, and let $\theta^\vee \in \bar{\mathfrak{h}}$ be the coroot corresponding to the highest root. Let $h^\vee = \phi(\theta^\vee, \theta^\vee)$ be the *dual Coxeter number* and let $(x|y) = \phi(x, y)/2h^\vee$ be the *normalized invariant form* on $\bar{\mathfrak{g}}$. The *affine algebra* $\mathfrak{g}'$ *associated to* $\bar{\mathfrak{g}}$ (called also the affine algebra of type $X_\ell^{(1)}$, where X_ℓ is the type of $\bar{\mathfrak{g}}$) is constructed as follows. Let $\mathbb{C}[t, t^{-1}]$ be the algebra of Laurent polynomials in t, and let us view the loop algebra $\bar{\mathfrak{g}}[t, t^{-1}] = \mathbb{C}[t, t^{-1}] \otimes_{\mathbb{C}} \bar{\mathfrak{g}}$ as an (infinite-dimensional) Lie algebra over $\mathbb{C}$. Then

$$\mathfrak{g}' = \bar{\mathfrak{g}}[t, t^{-1}] + \mathbb{C}K$$

is the unique non-trivial central extension of $\bar{\mathfrak{g}}[t, t^{-1}]$ by a 1-dimensional center $\mathbb{C}K$. Explicitly, it can be defined by the following commutation relations:

$$[x(m), y(n)] = [x, y](m + n) + m\delta_{m,-n}(x|y)K,$$

where $x(m) \in \bar{\mathfrak{g}}[t, t^{-1}]$ stands for $t^m \otimes x$ $(m \in \mathbb{Z},\ x \in \bar{\mathfrak{g}})$. We identify $\bar{\mathfrak{g}}$ with the subalgebra $1 \otimes \bar{\mathfrak{g}}$ of $\mathfrak{g}'$, and let $\mathfrak{h}' = \bar{\mathfrak{h}} + \mathbb{C}K$ be the Cartan subalgebra of $\mathfrak{g}'$. Let also $\mathfrak{n}_\pm = \bar{\mathfrak{n}}_\pm + t^{\pm 1}\bar{\mathfrak{g}}[t^{\pm 1}]$. Then we have the *triangular decomposition* $\mathfrak{g}' = \mathfrak{n}_- + \mathfrak{h}' + \mathfrak{n}_+$.

Supported in part by NSF grant DMS–8802489 and Sloan grant 88–10–1.
AMS subject classification (1980). 20C35.
Key words. Affine algebras, branching functions.

An arbitrary affine algebra is a direct sum of the Lie algebras of the form $\mathfrak{g}'$ and their twisted analogues $\mathfrak{g}'(\sigma, s)$ (see below).

It is often convenient to consider $\mathfrak{g}'$ as an ideal of codimension 1 in the Lie algebra $\mathfrak{g} = \mathfrak{g}' + \mathbb{C}d$, with commutation relations

$$[d, x(m)] = mx(m), \ [d, K] = 0,$$

and let $\mathfrak{h} = \mathfrak{h}' + \mathbb{C}d$ be its Cartan subalgebra. One extends the normalized bilinear form $(.|.)$ from $\overline{\mathfrak{h}}$ to $\mathfrak{h}$ by letting $(\overline{\mathfrak{h}}|\mathbb{C}K + \mathbb{C}d) = 0, \ (K|K) = (d|d) = 0, \ (K|d) = 1$.

A representation π of the affine algebra $\mathfrak{g}'$ in a vector space V is called a *positive energy* representation if a) $\pi(K) = kI, \ k \in \mathbb{C}$, and b) π can be extended to the whole $\mathfrak{g}$ such that $-\pi(d)$ is diagonalizable and its eigenvalues are non–negative integers.

The number k is called the *level* of V. The eigenspace decomposition

$$V = \oplus_{n \in \mathbb{Z}_+} V_{(n)}$$

with respect to $-\pi(d)$ is called the *energy decomposition*; if $v \in V_{(n)}$ we say that v has energy n.

Since $[d, \overline{\mathfrak{g}}] = 0$, the energy decomposition is $\overline{\mathfrak{g}}$–invariant, and we denote by $\overline{\pi}$ the representation of $\overline{\mathfrak{g}}$ in $V_{(0)}$. It is easy to show that the map $\pi \longmapsto (k, \overline{\pi})$ establishes a bijective correspondence between the set of (equivalence classes of) positive energy irreducible representations (π, V) such that $V_{(0)} \neq 0$ of $\mathfrak{g}'$ and the set of pairs $(k, \overline{\pi})$, where $k \in \mathbb{C}$ and $\overline{\pi}$ is an irreducible representation of $\overline{\mathfrak{g}}$ (considered up to equivalence). (Given $k \in \mathbb{C}$ and an irreducible representation $\overline{\pi}$ in $V_{(0)}$, we extend $\overline{\pi}$ to $\mathfrak{g}_+ := \overline{\mathfrak{g}}[t] + \mathbb{C}K + \mathbb{C}d$ by letting $\overline{\pi}(K) = kI_{V_{(0)}}, \ \overline{\pi}(d) = 0$ and $\overline{\pi}(x(n)) = 0$ for $n > 0$, and let π be the irreducible quotient of the induced representation of $\mathfrak{g}$ in the space $\mathcal{U}(\mathfrak{g}) \otimes_{\mathcal{U}(\mathfrak{g}_+)} V_{(0)}$. This gives us an irreducible positive energy representation π of $\mathfrak{g}'$ corresponding to the pair $(k, \overline{\pi})$.)

0.3. We shall consider only the irreducible positive energy representations (π, V) of $\mathfrak{g}'$ such that $(\overline{\pi}, V_{(0)})$ is an irreducible highest weight representation of $\overline{\mathfrak{g}}$. In other words, we shall assume that there exists a non–zero vector $v_{k,\overline{\lambda}} \in V_{(0)}$, where $\overline{\lambda} \in \overline{\mathfrak{h}}$, such that

$$\pi(\overline{\mathfrak{n}}_+)v_{k,\overline{\lambda}} = 0, \ \pi(h)v_{k,\overline{\lambda}} = \overline{\lambda}(h)v_{k,\overline{\lambda}} \text{ for } h \in \overline{\mathfrak{h}}.$$

These representations are parameterized by the pairs $(k, \overline{\lambda}), \ k \in \mathbb{R}, \ \overline{\lambda} \in \overline{\mathfrak{h}}^*$. It is more convenient to represent the pair $(k, \overline{\lambda})$ by an element $\lambda \in \mathfrak{h}'^*$ such that $\lambda|_{\overline{\mathfrak{h}}} = \overline{\lambda}$ and $\lambda(K) = k$. The corresponding representation π_λ of $\mathfrak{g}'$ is denoted by $L(\lambda)$ and is called the *irreducible highest weight representation* of $\mathfrak{g}'$ with *highest weight* λ.

The vector $v_\lambda := v_{k,\overline{\lambda}}$ is called the highest weight vector; it is the unique up to a non–zero constant factor vector in $L(\lambda)$ satisfying equations

$$\pi_\lambda(\mathfrak{n}_+)v_\lambda = 0, \ \pi_\lambda(h)v_\lambda = \lambda(h)v_\lambda \text{ for } h \in \mathfrak{h}'.$$

Especially important are the representations $L(\lambda)$ of $\mathfrak{g}'$ which can be lifted to a (projective) representation of the corresponding loop group. These are called *integrable* highest weight representation. They are parameterized by the set

$$P_+ = \{\lambda \in \mathfrak{h}'^* | k := \lambda(K) \in \mathbb{Z}_+, \ \overline{\lambda} \in \overline{P}_+, \ (\overline{\lambda}|\theta^\vee) \leq k\}.$$

Here $\overline{P}_+ \subset \overline{\mathfrak{h}}^*$ is the set of highest weights of finite–dimensional irreducible representations of $\overline{\mathfrak{g}}$.

The basic tool for study of integrable $L(\lambda)$ is the so called Weyl–Kac character formula [6] for the function ch_λ on $Y := \{v \in \mathfrak{h} | \mathrm{Re}(v|K) > 0\}$ defined by

$$\mathrm{ch}_\lambda(v) = \mathrm{tr}_{L(\lambda)} e^v.$$

Recently a similar character formula has been established for a larger class of the $L(\lambda)$, those with an "admissible" highest weight [12]. Their "normalized" characters are Jacobi modular forms and, conjecturally, these representations are characterized by this property. (These are also the only ones for which the Kazhdan–Lusztig polynomials are trivial.) In the present paper we consider the best studied (see [13]), principal admissible highest weight representations. Their levels may be arbitrary rational numbers k (called principal admissible) such that $k + h^\vee \geq h^\vee/u$, where $u \in \mathbb{N}$ is the denominator of k. In the case when $k \in \mathbb{Z}_+$ all principal admissible representations are integrable (but all representations of fractional level are not).

0.4. A natural class of subalgebras of the Lie algebra $\mathfrak{g}'$ (and similarly of an arbitrary affine algebra) to consider is the following. Let $\dot{\overline{\mathfrak{g}}}$ be a reductive subalgebra of $\overline{\mathfrak{g}}$, let σ be an automorphism of $\dot{\overline{g}}$ and let $s \in \mathbb{N}$ be such that $\sigma^s = 1$. Define an automorphism $\tilde{\sigma}$ of the subalgebra $\dot{\overline{\mathfrak{g}}}[t, t^{-1}] + \mathbb{C}K$ of $\mathfrak{g}'$ by letting

$$\tilde{\sigma}(x(n)) = (\exp \frac{2\pi i n}{s})\sigma(x)(n), \ \sigma(K) = K,$$

and denote by $\dot{\mathfrak{g}}'(\sigma, s)$ the fixed point set of $\tilde{\sigma}$.

0.5. We shall consider only the representations $L(\lambda)$ of $\mathfrak{g}'$ which are completely reducible with respect to $\dot{\mathfrak{g}}'(\sigma, s)$. (This is always the case when $\lambda \in P_+$.) The branching coefficients of such $\dot{\mathfrak{g}}'(\sigma, s)$ in a representation $L(\lambda)$ of $\mathfrak{g}'$, i.e. the numbers $[L(\lambda) : \dot{L}(\mu)]$, where $\dot{L}(\mu)$ is an irreducible highest weight representation of $\dot{\mathfrak{g}}(\sigma, s)$, are almost always zero or infinity. To get around this, let $\dot{\mathfrak{g}}(\sigma, s) = \dot{\mathfrak{n}}_- + \dot{\mathfrak{h}} + \dot{\mathfrak{n}}_+$ be the induced triangular decomposition of $\dot{\mathfrak{g}}(\sigma, s)$, i.e. $\dot{\mathfrak{n}}_\pm = \mathfrak{n}_\pm \cap \dot{\mathfrak{g}}(\sigma, s)$ and $\dot{\mathfrak{h}}' = \mathfrak{h}' \cap \dot{\mathfrak{g}}(\sigma, s)$, and let

$$[\lambda : \mu]_n = \dim\{v \in L(\lambda)_{(n)} | \pi_\lambda(\dot{\mathfrak{n}}_+)v = 0, \ \pi_\lambda(h)v = \mu(h)v \text{ for all } h \in \dot{\mathfrak{h}}'\}.$$

These numbers are always finite, and we can consider the power series

$$b_\mu^\lambda = q^{m_{\lambda,\mu}} \sum_{n \in \mathbb{Z}_+} [\lambda : \mu]_n q^n.$$

The number $m_{\lambda,\mu}$ is a rational number called the *modular anomaly* which is defined as follows. Let $q = e^{2\pi i \tau}$; then the above series converges to a holomorphic function in τ for $\mathrm{Im}\,\tau > 0$. According to [10, Proposition 4.36], provided that $\lambda \in P_+$, there exists a unique number $m_{\lambda,\mu}$ such that the function $b_\mu^\lambda(\tau)$ is a modular function, i.e. is fixed under the action of a principal congruence subgroup $\Gamma(N)$, some N. (Explicit formulas for the $m_{\lambda,\mu}$ may be found in [11] and in the paper; of course, these numbers depend also of the $\mathfrak{g}'$ and the subalgebra.)

The functions $b^\lambda_\mu(\tau)$, clearly, completely describe the decomposition of $L(\lambda)$ with respect to the subalgebra in question. They are called *branching functions.* If $[L(\lambda) : \dot{L}(\mu)] < \infty$, then, of course, $[L(\lambda) : \dot{L}(\mu)] = \lim_{\tau\downarrow 0} b^\lambda_\mu(\tau)$. In general, we study the asymptotics of $b^\lambda_\mu(\tau)$ as $\tau \downarrow 0$ instead. Namely, since the $b^\lambda_\mu(\tau)$ are modular functions, we have as $\tau \downarrow 0$, provided that $b^\lambda_\mu \neq 0$:

$$b^\lambda_\mu(\tau) \sim a(\lambda,\mu) e^{\frac{\pi i}{12\tau} g(\lambda,\mu)},$$

where $a(\lambda,\mu) > 0$ and $g(\lambda,\mu) \geq 0$ are real numbers called the *asymptotic dimension* and *growth* of the branching function. Here and in the rest of the paper $f(\tau) \sim g(\tau)$ means that $\lim_{\tau\downarrow 0} f(\tau)/g(\tau) = 1$.

In all known examples the growth depends only on the algebra $\mathfrak{g}'$, the subalgebra and the level k of λ, but this is an open problem, which we will refer to as the basic conjecture, even in the case of integrable $L(\lambda)$ and the subalgebra $\dot{\mathfrak{g}}(1,1)$ (cf. [11]). Incidentally, the knowledge of the above asymptotics allows one to compute the asymptotics of the branching coefficients $[\lambda : \mu]_n$ as $n \to \infty$ by making use of a Tauberian theorem (see [10] and [11]).

0.6. An important version of a special case of branching functions are *string functions* c^λ_μ, which correspond to $\dot{\overline{\mathfrak{g}}} = \overline{\mathfrak{h}}$, $\sigma = 1$, $s = 1$, i.e. which describe the multiplicities of weights of $L(\lambda)$. Namely, given $\lambda, \mu \in \mathfrak{h}'^*$ of level k, we define as usual, the weight space $L(\lambda)_\mu = \{v \in L(\lambda) | h(v) = \mu(h)v \text{ for all } h \in \mathfrak{h}'\}$, and let

$$c^\lambda_\mu = q^{m_{\lambda,\mu}} \sum_{n\in\mathbb{Z}_+} (\dim L(\lambda)_\mu \cap L(\lambda)_{(n)}) q^n,$$

where we let $m_\lambda = |\overline{\lambda} + \overline{\rho}|^2/2(k+h^\vee) - |\overline{\rho}|^2/2h^\vee$ to be the modular anomaly of λ (see below) and $m_{\lambda,\mu} = m_\lambda - |\overline{\mu}|^2/2k$ is the modular anomaly in this case (as usual, $\overline{\rho}$ is the half-sum of positive roots for $\overline{\mathfrak{g}}$). This is a modular form of weight $-\ell/2$, and it is related to the corresponding branching function by the equation

$$c^\lambda_\mu(\tau) = G(\tau)^{-1} b^\lambda_\mu(\tau),$$

where $G(\tau)$ is a modular form of weight $\ell/2$ given in §2.2.

String functions for integrable $L(\lambda)$ were studied in great detail in [9], [10], [11] (see also [8, Chapters 12 and 13]). The key result of this work is an explicit transformation formula for the normalized characters

$$\chi_\lambda := q^{m_\lambda} \mathrm{ch}_\lambda,\ \lambda \in P_+,$$

under the action of the involution $S : \tau \longmapsto -1/\tau$ (defined in §4.3). Here we identify q with the function $e^{-K}(v) = e^{-(K|v)}$ on $\mathfrak{h}$. (This result was extended in [13, Theorem 3.6] to the case of principal admissible $L(\lambda)$; see formula (4.3.1) of the present paper.) One deduces from this result an explicit transformation formula for string functions under the involution S [10, Theorem A]. Since the transformation formula together with the polar parts of the q-expansions completely determine modular forms, this allows one to compute the string functions explicitly in many interesting cases. Furthermore, it turns out that the asymptotics of the string functions c_μ^λ is independent of μ, which allows one to find this asymptotics explicitly ([10, Proposition 4.21] or [8, Chapter 13],) proving thereby the validity of the basic conjecture in this case. (Incidentally, for admissible λ the string functions $c_\mu^\lambda(\tau)$ fail to be modular forms [14].)

0.7. General branching functions in the case of the subalgebras of the form $\dot{\mathfrak{g}}'(1,1) = \dot{\overline{\mathfrak{g}}}[t,t^{-1}] \oplus \mathbf{C}K$ of $\mathfrak{g}'$ and integrable $L(\lambda)$ were studied in detail in [11]. Again from the tranformation formula for the normalized characters one deduces the transformation law for these branching functions under the involution S [11, Theorem A]. One derives from this definitive results on asymptotics (which prove the validity of the basic conjecture) only in the case of tensor products [11, (2.7.15) and §3.4]. For general branching functions we derive the basic conjecture from the conjectural positivity of certain matrix elements of the transformations S in the basis of branching functions [11, p. 188].

0.8. In the present paper we consider the subalgebras

$$\mathfrak{g}'_{[u]} := \mathfrak{g}'(1,u) = \overline{\mathfrak{g}}[t^u, t^{-u}] \oplus \mathbf{C}K$$

of $\mathfrak{g}'$, called the *winding* subalgebras. (This is the simplest case different from $\dot{\mathfrak{g}}'(1,1)$.) The first basic result of the paper is Theorem 2.1 which gives an explicit expression of the branching functions b_λ^Λ for winding subalgebras in integrable highest weight representations $L(\Lambda)$ in terms of string functions. The special case of this theorem, when the level of $L(\Lambda)$ is 1, is Theorem 2.2, which gives a solution to Frenkel's conjecture [4]. Theorem 2.1 leads also to Conjecture 2.2 on the asymptotics of the branching functions for winding subalgebras. We were able to prove only that it holds for all sufficiently large u.

0.9. Next we compare branching functions b_λ^Λ for winding subalgebras with branching functions $b_\lambda^{\Lambda\otimes\mu}$ for tensor products $L(\Lambda)\otimes L(\mu)$. It is clear from [13, Corollary 4.1 and Theorem 3.6] that the branching functions $b_\lambda^{\Lambda\otimes\mu}$ are modular functions provided that Λ is integrable and μ is admissible. This is the case studied in the present paper. As in the case of winding subalgebras, we find an explicit expression of the $b_\lambda^{\Lambda\otimes\mu}$ in terms of string functions (Theorem 3.1). Comparing Theorems 2.1 and 3.1, we see that a branching function b_λ^Λ for a winding subalgebra $\mathfrak{g}'_{[u]}$ coincides with a branching function for tensor product $L(\Lambda)\otimes L(\mu)$, where $\mu = ((u^{-1}-1)h^\vee, 0)$, provided that u is relatively prime to $h^\vee$ and $r^\vee$ (Proposition 3.2). This (still mysterious) coincidence indicates a remarkable interplay between the integrable and admissible representations. As in the case of winding subalgebras, Theorem 3.1 leads to Conjecture 3.1 on asymptotics of the $b_\lambda^{\Lambda\otimes\mu}$, generalizing the known result in the integrable case. After this paper was completed we received preprint [1] where Theorem 3.1 is derived in the integrable case using a free field resolution.

0.10. The remarkable feature of the theory of integrable, and, more generally, principal admissible highest weight representations is the $SL_2(\mathbf{Z})$–invariance of the $\mathbf{C}$–span of normalized characters (in the twisted case, $SL_2(\mathbf{Z})$ should be replaced by a slightly smaller subgroup Γ; see Proposition 4.3), hence the $SL_2(\mathbf{Z})$–invariance of the $\mathbf{C}$–span of branching functions $b_\lambda^{\Lambda\otimes\mu}$, where Λ (resp. μ) runs over all integrable (resp. principal admissible) weights of fixed level. (A similar result holds for arbitrary subalgebras of the form $\dot{\mathfrak{g}}'(1,1)$, see [11]. This follows from the S–invariance of normalized characters.) Due to the above coincidence, this is the case also for the branching functions for winding subalgebras $\mathfrak{g}'_{[u]}$

provided that u is relatively prime to $h^\vee$ and $r^\vee$. However, in the general case we have only the $\Gamma_0(u)$–invariance (see §4.3). This is a general feature of the subalgebras $\dot{\mathfrak{g}}'(\sigma, s)$.

0.11. The last, Section 4, contains some preparatory material for our forthcoming paper with E. Frenkel [3]. It deals with functions $\varphi_{\lambda,\mu}$, which are branching functions for tensor products of the level 1 integrable representations with arbitrary principal admissible representations (see (4.1.1) and Theorem 4.1). These functions previously appeared in this context in [12, Proposition 3].

It turns out that the functions $\varphi_{\lambda,\mu}$ can be obtained by a simple limiting procedure from the characters of the principal admissible representations (Proposition 4.2). As will be explained in [3] this procedure naturally appears in the quantization of the Drinfeld–Sokolov reduction developed in [2]. As a result, one obtains that the functions $\varphi_{\lambda,\mu}$ are characters of the so called extended conformal algebras, which are higher rank generalizations of the Virasoro algebra (cf. [15]). In the particular case of $\bar{\mathfrak{g}} = s\ell_2(\mathbb{C})$, this procedure is equivalent to taking the residue of the admissible characters. As was shown previously in [16], this reproduces the Virasoro characters. It is worth mentioning that the limiting procedure gives a non–zero result only for "non–degenerate" principal admissible weights; in particular, the integrable characters always give zero.

The main result of this last section is Theorem 4.4 which give a transformation formula for the $\varphi_{\lambda,\mu}$ under the action of S, obtained from the limiting procedure (which is simpler than that obtained from tensor products). This formula will be applied in [3] to calculate the fusion rules for the extended conformal algebras.

Thus Theorem 4.1b means the coincidence of two theories of extended conformal algebras at least on the character level in the simply laced or twisted case. (In the case of $B_\ell^{(1)}$ they are different, as can be seen by comparing Theorem 2.2′ and Proposition 4.2.)

Note that though the set of principal admissible representations of given fractional level carries quite a few features of a conformal field theory (like modular invariance, the unique vacuum, the involution), it can't be a conformal field theory since, for example, its fusion rules computed by Verlinde's formula [17] may be negative. This makes it quite remarkable that a "reduction" of this theory indeed produces a conformal field theory.

0.12. We would like to thank E. Getzler and M. Hopkins who pointed out that the study of branching functions for winding subalgebras may be important for the theory of cohomological operations in the elliptic cohomology, which stimulated our research. We thank E. Frenkel for his patient explanations of his (joint with Feigin) work [2] and for collaboration in Section 4. We thank D. Jerison for consultations on asymptotics.

The first author wishes to thank E. Strickland for the invitation to give a talk in January 1990 in the Universita di Roma II on the subject of this paper, for her persistence in having the paper written and for her lavish hospitality. The second author acknowledges the hospitality of MIT, where during his stay in the spring of 1990 the paper was completed.

§1. Notation and preliminaries.

1.1. Let $I = \{0, 1, \ldots, \ell\}$, $\ell \geq 1$. Recall that an *affine matrix* is a square matrix $A = (a_{ij})_{i,j \in I}$ such that $a_{ii} = 2$, $-a_{ij} \in \mathbb{Z}_+$ for $i \neq j$, $a_{ij} = 0$ implies $a_{ji} = 0$, and there exists a unique sequence $(a_0, \ldots, a_\ell)$ of positive relatively prime integers, called the *null-vector* of A, such that $(a_0, \ldots, a_\ell)({}^tA) = 0$.

Two affine matrices are called *equivalent* if they are obtained from each other by a reordering of I. A complete list of affine matrices up to equivalence is given in Tables Aff r of [8, Chapter 4]. Affine matrices in these tables are denoted by their type $X_N^{(r)}$, where X_N is the type of the "underlying" simple finite–dimensional Lie algebra and r is the so–called *tier number*. The number $h = \sum_{i\in I} a_i$ is called the *Coxeter number* of A.

Note that the transposed matrix ${}^tA$ is again affine. The corresponding null–vector is denoted by $(a_0^\vee, \dots, a_\ell^\vee)$, the tier number by $r^\vee$ and the Coxeter number by $h^\vee$. The latter is called the *dual Coxeter number* of A.

Let $\varepsilon_j = a_j^\vee a_j^{-1}$; then the matrix $(\varepsilon_i a_{ij})_{i,j\in I}$ is symmetric.

We keep the ordering of I given in [8, Chapter 4]. For this ordering we have in particular: $a_0^\vee = 1$ and $a_0 = 1$ (resp. $a_0 = 2$) if $A \neq A_{2\ell}^{(2)}$ (resp. $A = A_{2\ell}^{(2)}$).

In what follows we fix an affine matrix A of type $X_N^{(r)}$.

1.2. Let $\mathfrak{h}'$ be the $\ell+1$–dimensional vector space over $\mathbb{C}$ with a basis $\Pi = \{\alpha_0, \dots, \alpha_\ell\}$ and a symmetric bilinear form defined in this basis by the following formula [8, Chapter 6]

$$(\alpha_i|\alpha_j) = \varepsilon_i a_{ij},\ i,j \in I. \tag{1.2.1}$$

Then Π is called a *root basis* and $(.|.)$ is called the *normalized invariant form* (for the matrix A). The basis $\Pi^\vee = \{\alpha_0^\vee, \dots, \alpha_\ell^\vee\}$ defined by $a_i\alpha_i = a_i^\vee \alpha_i^\vee$ is called a *coroot basis.* Note that the kernel of the normalized invariant form consists of all multiples of the element

$$K = \sum_{i\in I} a_i\alpha_i = \sum_{i\in I} a_i^\vee \alpha_i^\vee, \tag{1.2.2}$$

called the *canonical central element.*

Given a subset L of $\mathfrak{h}'$, we let $\mathbb{Z}L$ (resp. $\mathbb{Z}_+L$, $\mathbb{C}L$, etc.) denote the set of all linear combinations of elements from L with coefficients from $\mathbb{Z}$ (resp. $\mathbb{Z}_+$, $\mathbb{C}$, etc.). We also let

$$L^* = \{\alpha \in \mathbb{C}L | (\alpha|L) \subset \mathbb{Z}\}.$$

The lattices $Q = \mathbb{Z}\Pi$ and $Q^\vee = \mathbb{Z}\Pi^\vee$ are called the *root* and *coroot* lattices respectively. We also let $Q_+ = \mathbb{Z}_+\Pi$, $Q_+^\vee = \mathbb{Z}_+\Pi^\vee$.

We let $\overline{I} = \{1, \dots, \ell\} \subset I$, $\overline{\Pi} = \{\alpha_1, \dots, \alpha_\ell\}$, $\overline{\Pi}^\vee = \{\alpha_1^\vee, \dots, \alpha_\ell^\vee\}, \overline{Q} = \mathbb{Z}\overline{\Pi}$, $\overline{Q}^\vee = \mathbb{Z}\overline{\Pi}^\vee, \overline{\mathfrak{h}} = \mathbb{C}\overline{\Pi}$. One has [8, Chapter 6]:

$$\overline{Q} \supset \overline{Q}^\vee \text{ if } r = 1;\ \overline{Q} \subset \overline{Q}^\vee \text{ if } r > 1.$$

Define the following important lattices M and $\tilde{M}$ by

$$M = \overline{Q} \text{ if } r^\vee = 1;\ M = \overline{Q}^\vee \text{ if } r^\vee > 1.$$
$$\tilde{M} = \overline{Q}^* \text{ if } r = 1;\ \tilde{M} = \overline{Q}^{\vee *} \text{ if } r > 1.$$

Since $(\overline{Q}|\overline{Q}^\vee) \subset \mathbb{Z}$, we have $M \subset \tilde{M} \subset M^*$. We also have:

(1.2.3) $$|M^*/\tilde{M}| \text{ (resp. } |\tilde{M}/M|) \text{ are the same for } A \text{ and } {}^tA,$$

(1.2.4) $$|M^*/\tilde{M}| = |\overline{Q}^\vee/M| \text{ if } r > 1.$$

1.3. Let $r_i \in \mathrm{Aut}\mathfrak{h}'$ be the *fundamental reflections*, i.e.

$$r_i(v) = v - (\alpha_i|v)\alpha_i^\vee,$$

and let $W =< r_i | i \in I >$ be the Weyl group. Note that Q and $Q^\vee$ are W-invariant.

Let $\Delta^{re} = W(\Pi)$, and $\Delta^{\vee re} = W(\Pi^\vee)$ be the sets of *real roots* and *real coroots* respectively. Let $\Delta_+^{re} = \Delta^{re} \cap Q_+$, $\Delta_+^{\vee re} = \Delta^{\vee re} \cap Q_+^\vee$. We also let $\overline{\Delta} = \Delta^{re} \cap \overline{Q}$, $\overline{\Delta}^\vee = \Delta^{\vee re} \cap \overline{Q}^\vee$, $\overline{\Delta}_+ = \Delta_+^{re} \cap \overline{Q}$, etc.

If $\alpha = w(\alpha_i) \in \Delta^{re}$, then $\alpha^\vee = w(\alpha_i^\vee) \in \Delta^{\vee re}$ is well-defined [8, Chapter 3] and $\alpha^\vee = 2\alpha/(\alpha|\alpha)$; letting $r_\alpha = w r_i w^{-1}$, we have:

(1.3.1) $$r_\alpha(v) = v - (\alpha|v)\alpha^\vee, \ v \in \mathfrak{h}'.$$

We have the following homomorphism $\alpha \longmapsto t_\alpha$ of $\mathfrak{h}'$ into $\mathrm{Aut}\mathfrak{h}'$ with kernel $\mathbb{C}K$ [8, Chapter 6]:

(1.3.2) $$t_\alpha(v) = v + (v|K)\alpha - ((v|\alpha) + \tfrac{1}{2}(\alpha|\alpha)(v|K))K.$$

For a subset L of $\mathfrak{h}'$ we let $t_L = \{t_\alpha | \alpha \in L\}$.

We have [8, Chapter 6]:

(1.3.3) $$W = t_M \rtimes \overline{W},$$

where $\overline{W} =< r_i | i \in \overline{I} >$ is a finite subgroup of W (we shall write: $w = t_\alpha \overline{w}$).

The sets of real roots and coroots are invariant with respect to the group $\tilde{W} := t_{\tilde{M}} \rtimes \overline{W}$ containing W. Let $\tilde{W}_+ = \{w \in \tilde{W} | w(\Pi) = \Pi\}$, and let

$$J = \{j \in I | a_j = 1 \text{ (resp. } a_j^\vee = 1) \text{ if } r = 1 \text{ (resp. } r > 1)\}.$$

Then [10, §4.8] the group $\tilde{W}_+$ is isomorphic to $\tilde{M}/M$, acts simply transitively on $\{\alpha_i \in \Pi | i \in J\}$, and one has:

(1.3.4) $$\tilde{W} = \tilde{W}_+ \ltimes W.$$

The group $\tilde{W}_+$ can be described more explicitly as follows. Let $\overline{W}_+$ be the subgroup of $\overline{W}$ consisting of the elements preserving the subset $\{\overline{\alpha}_0 := a_0\alpha_0 - K, \overline{\alpha}_1 := \alpha_1, \ldots, \overline{\alpha}_\ell := \alpha_\ell\}$

of $\overline{Q}$. This group acts simply transitively on the set $\{\overline{\alpha}_j | j \in J\}$. Denote by $\overline{w}_j$ the element of $\overline{W}_+$ such that $\overline{w}_j(\overline{\alpha}_0) = \overline{\alpha}_j$. Then

$$w_j := t_{\Lambda_j - \Lambda_0} \overline{w}_j$$

is the element of $\tilde{W}_+$ such that $w_j(\alpha_0) = \alpha_j$.

Note also the following useful formulas:

$$|\tilde{M}/M| = |J|, \tag{1.3.5}$$

$$|M^*/M| = |J| \text{ if } r = r^\vee = 1 \text{ or } a_0 = 2. \tag{1.3.6}$$

1.4. Define the *fundamental weights* (resp. *fundamental coweights*) $\Lambda_i \in \mathfrak{h}'^*$ (resp. $\Lambda_i^\vee \in \mathfrak{h}'^*$), $i \in I$, by:

$$< \Lambda_i, \alpha_j^\vee > \text{(resp. } < \Lambda_i^\vee, \alpha_j >) = \delta_{ij},\ j \in I.$$

Let $P = \sum_{i\in I} \mathbb{Z}\Lambda_i$, $P^\vee = \sum_{i\in I} \mathbb{Z}\Lambda_i^\vee$ be the sets of integral weights and coweights and $P_+ = \sum_{i\in I} \mathbb{Z}_+\Lambda_i$, $P_+^\vee = \sum_{i\in I} \mathbb{Z}_+\Lambda_i^\vee$ the sets of dominant integral weights. Let

$$\rho = \sum_{i\in I} \Lambda_i,\ \rho^\vee = \sum_{i\in I} \Lambda_i^\vee.$$

Note that

$$< \rho^\vee, K >= h,\ < \rho, K >= h^\vee. \tag{1.4.1}$$

The number $k =< \lambda, K >$ is called the *level* of $\lambda \in \mathfrak{h}'^*$; the set of all λ of level k is denoted by $\mathfrak{h}'^{*k}$. The following map τ_k sends $\mathfrak{h}'^*$ to $\mathfrak{h}'^{*k}$ (since $< \Lambda_0, K >= 1$):

$$\tau_k(\lambda) = \lambda - (< \lambda, K > -k)\Lambda_0. \tag{1.4.2}$$

Fix $u \in \mathbb{N}$ relatively prime to a_0. It is easy to see from [8, Chapter 6] that the element $\dot{\alpha}_0 := a_0^{-1}(u-1)K + \alpha_0$ lies in Δ_+^{re} and the element $\dot{\alpha}_0^\vee := (u-1)K + \alpha_0^\vee$ lies in $\Delta_+^{\vee re}$. We let

$$\begin{aligned}
\Pi_{[u]} &= \{\dot{\alpha}_0, \dot{\alpha}_i := \alpha_i\ (i \in \overline{I})\},\\
\Pi_{[u]}^\vee &= \{\dot{\alpha}_0^\vee, \dot{\alpha}_i^\vee := \alpha_i^\vee\ (i \in \overline{I})\},\\
W_{[u]} &=< r_{\dot{\alpha}_i} | i \in I > .
\end{aligned}$$

Then $\Pi_{[u]}$ and $\Pi_{[u]}^\vee$ are root and coroot bases with the same normalized bilinear form as for Π and $\Pi^\vee$. We shall indicate the objects associated to $\Pi_{[u]}$ by an overdot. For example, the canonical central element for $\Pi_{[u]}$ is

(1.4.3)
$$\dot{K} := \sum_{i \in I} a_i \dot{\alpha}_i^\vee = uK,$$

the fundamental weights are

(1.4.4)
$$\dot{\Lambda}_i = \Lambda_i + (u^{-1} - 1) a_i^\vee \Lambda_0,$$

hence

(1.4.5)
$$\dot{\rho} := \sum_{i \in I} \dot{\Lambda}_i = \rho + (u^{-1} - 1) h^\vee \Lambda_0, \ |\dot{\rho}|^2 = |\rho|^2,$$

(1.4.6)
$$\dot{t}_\alpha = t_{u\alpha},$$

(1.4.7)
$$\dot{W} = W_{[u]} = t_{uM} \rtimes \overline{W}.$$

The following lemma, the proof of which is straightforward, will be useful in the sequel.

LEMMA 1.4. *If* $y \in \tilde{W}$, $\gamma \in \overline{\mathfrak{h}}^*$ *and* $w \in yW_{[u]}y^{-1}$, *so that* $y = t_\beta \overline{y}$ *for some* $\beta \in \tilde{M}$ *and* $y \in \overline{W}$, *and* $w = yt_{u\alpha}\overline{w}y^{-1}$ *for some* $\alpha \in M$ *and* $\overline{w} \in \overline{W}$, *then*

$$wyt_\gamma y^{-1} w^{-1} = t_{\overline{w}\ \overline{y}(\gamma)}. \quad \square$$

1.5. Let $k \in \mathbb{Q}$ and let $u \in \mathbb{N}$ be the denominator of k (i.e. $ku \in \mathbb{Z}$ and $(ku, u) = 1$). We say that k is *principal admissible* if

(1.5.1)
$$u(k + h^\vee) \geq h^\vee \ \text{ and } \ (u, r^\vee) = 1.$$

Given $k \in \mathbb{Z}_+$ we let $P_+^k = P_+ \cap \mathfrak{h}'^{*k}$. Given a principal admissible k, we let $P_+^k = \tau_k(P^{u(k+h^\vee)-h^\vee})$. (Note that when $k \in \mathbb{Z}_+$, the two definitions of P_+^k coincide.)

Given $\lambda \in \mathfrak{h}'^*$, we let

$$R^\lambda = \{\alpha \in \Delta^{\vee re} | < \lambda + \rho, \alpha > \in \mathbb{Z}\}, \ R_+^\lambda = R^\lambda \cap \Delta_+^{\vee re},$$

and denote by S^λ the set of $\alpha \in R_+^\lambda$ which do not decompose into a sum of several elements from R_+^λ. Let $W^\lambda = < r_\alpha | \alpha \in R_+^\lambda >$; one can show that $W^\lambda = < r_\alpha | \alpha \in S^\lambda >$.

We call $\lambda \in \mathfrak{h}'^*$ an *admissible weight* if it satisfies the following two properties:

(1.5.2)
$$- < \lambda + \rho, \alpha > \notin \mathbb{Z}_+ \text{ for all } \alpha \in \Delta_+^{\vee re},$$

$$\mathbb{Q}R^\lambda = \mathbb{Q}\Pi^\vee. \tag{1.5.3}$$

A weight λ is called *principal admissible* if in addition to (1.5.2 and 3) we have:

$$\text{the matrix } (2(\alpha|\beta)/(\beta|\beta))_{\alpha,\beta\in S^\lambda} \text{ is equivalent to } A. \tag{1.5.4}$$

Note that all dominant integral weights are principal admissible.

We recall now the description of the set of all principal admissible weights. Given $y \in \tilde{W}$, denote by $P_{u,y}$ the set of all admissible λ such that $S^\lambda = y(\Pi^\vee_{[u]})$; let $P^k_{u,y} = P_{u,y} \cap \mathfrak{h}'^{*k}$. Denote by $P^k(A)$ the set of all principal admissible weights of level k. Finally, recall the shifted action of $\tilde{W}$:

$$w.\lambda = w(\lambda + \rho) - \rho.$$

PROPOSITION 1.5. *[13, Theorem 2.1 and Proposition 2.1].*
(a) $P^k_{u,y} \neq \emptyset$ if and only if

$$k \text{ is principal admissible and } u \text{ is the denominator of } k, \tag{1.5.5}$$

$$y(\Pi^\vee_{[u]}) \subset \Delta^{\vee re}_+. \tag{1.5.6}$$

(b) If (k,u,y) and (k,u,y') are two triples satisfying (1.5.5 and 6), then the following statements are equivalent:
(i) $P^k_{u,y} \cap P^k_{u,y'} \neq \emptyset$,
(ii) $P^k_{u,y} = P^k_{u,y'}$,
(iii) $y(\Pi^\vee_{[u]}) = y'(\Pi^\vee_{[u]})$,.
(iv) there exists $\sigma = t_\alpha\overline{\sigma} \in \tilde{W}_+$ such that $y' = yt_{u\alpha}\overline{\sigma}$.

(c) If (1.5.5 and 6) hold, then

$$P^k_{u,y} = y.P^k_+ = \{y.(\lambda - (u-1)(k+h^\vee)\Lambda_0) | \lambda \in P^{u(k+h^\vee)-h^\vee}_+\}.$$

(d) $P^k(A) \neq \emptyset$ if and only if k is principal admissible.
(e) $P^k(A) = \cup_y P^k_{u,y}$, where u is the denominator of k and y satisfies (1.5.6). □

PROOF: We shall prove that (iii) is equivalent to (iv). The rest of the statements are proved in [13]. Indeed, (iii) is equivalent to $y^{-1}y' \in \dot{\tilde{W}}_+$. But $t_\alpha\overline{\sigma} \in \tilde{W}_+$ if and only if $t_{u\alpha}\overline{\sigma} \in \dot{\tilde{W}}_+$. □

Remark 1.5. (a) If $k \in \mathbb{Z}_+$, then $P^k(A) = P^k_+$.

(b) Admissible weights are classified completely in [13]. In the case $A = A_\ell^{(1)}$ this is precisely the set $P(A)$ of all principal admissible weights. For all other affine matrices there are admissible weights which are not principal admissible, and their levels need not be principal admissible.

For $\lambda \in P^k_{u,y}$, let $\lambda^0 = \tau_k^{-1}(y^{-1}.\lambda)$, i.e.

$$\lambda = y.(\lambda^0 - (u-1)(k+h^\vee)\Lambda_0). \tag{1.5.7}$$

The map $\lambda \longmapsto \lambda^0$ defines a bijective correspondence between $P^k_{u,y}$ and $P_+^{u(k+h^\vee)-h^\vee}$.

Note that

$$W^\lambda = yW_{[u]}y^{-1} \text{ for } \lambda \in P^k_{u,y}. \tag{1.5.8}$$

The set $P^k(A)$ admits an important involution $\lambda \longmapsto {}^t\lambda$ defined as follows. Let $\lambda \in P^k_{u,y}$, $y = t_\beta \overline{y}$, $\beta \in \tilde{M}$, $\overline{y} \in \overline{W}$. First, note that $\overline{W}^\lambda := W^\lambda \cap \overline{W}$ is the Weyl group for the finite root system $\overline{\Delta}^\lambda := \overline{\Delta}^\vee \cap R^\lambda$. Denote by $\overline{w}^\lambda$ the longest element in $\overline{W}^\lambda$ (so that $\overline{w}^0$ is the longest element in $\overline{W}$). Recall that we have: $\overline{w}^\lambda \overline{\Delta}_+^\lambda = -\overline{\Delta}_+^\lambda$ and $(\overline{w}^\lambda)^2 = 1$. Define $w^\lambda \in \mathrm{Aut}\,\mathfrak{h}$ by

$$w^\lambda(v) = -\overline{w}^\lambda(v) \text{ if } v \in \overline{\mathfrak{h}},\ w^\lambda(K) = K,\ w^\lambda(d) = d. \tag{1.5.9}$$

Note that $w^\lambda \Delta_+^{\vee re} = \Delta_+^{\vee re}$. If $\Lambda \in P_+^m$ ($m \in \mathbb{Z}_+$) is a dominant integral weight, we let

$${}^t\Lambda = w^0(\Lambda)(= -\overline{w}^0(\Lambda) + 2m\Lambda_0 \in P_+^m). \tag{1.5.10}$$

Let

$${}^t\overline{y} = \overline{y}\overline{w}^\lambda\overline{w}^0,\ {}^ty = t_{-\beta}\,{}^t\overline{y}. \tag{1.5.11}$$

Finally, we let

$${}^t\lambda = {}^ty.({}^t(\lambda^0) - (u-1)(k+h^\vee)\Lambda_0). \tag{1.5.12}$$

It is easy to check that this is an involution that maps $P^k_{u,y}$ onto $P^k_{u,{}^ty}$, and that more invariantly this involution can be defined by

$${}^t\lambda = w^\lambda(\lambda+\rho) - \rho. \tag{1.5.13}$$

We clearly have

$$R^{{}^t\lambda} = w^\lambda(R^\lambda). \tag{1.5.14}$$

The following lemma will be used later.

LEMMA 1.5. *Let $y = t_\beta \overline{y} \in \tilde{W}$, where $\beta \in \tilde{M}$, $y \in \overline{W}$, and let $u \in \mathbb{N}$ be such that*

$$y(\Pi^\vee_{[u]}) \subset \Delta^{\vee re}_+ \backslash \overline{\Delta}^\vee_+. \tag{1.5.15}$$

Define $\omega_{u,y}$ by

$$\omega_{u,y} + \rho (resp. + \rho^\vee) = u\Lambda_0 - \overline{y}^{-1}(\beta) \text{ if } r = r^\vee = 1 \text{ or } r > 1 (resp. \text{ if } r = 1,\ r^\vee > 1). \tag{1.5.16}$$

Then $\omega_{u,y} \in P^{u-h^\vee}_+$ (resp. $\in P^{\vee u-h}_+$).

PROOF: We shall consider the case $r = r^\vee = 1$ or $r > 1$. The proof in the case $r = 1$, $r^\vee > 1$ is similar. It is clear that $\omega_{u,y} \in P^{u-h^\vee}$, so one has to show that $(\omega_{u,y}|\alpha_i^\vee) \geq 0$. For $i > 0$ we have: $0 < y(\dot{\alpha}_i^\vee) = \overline{y}\alpha_i^\vee - (\overline{y}\alpha_i^\vee|\beta)K$, hence $(\alpha_i^\vee|\overline{y}^{-1}\beta) < 0$ and $(\omega_{u,y}|\alpha_i^\vee) \geq 0$. Finally, $0 < y(\dot{\alpha}_0^\vee) = (u - 1 - (\alpha_0^\vee|\overline{y}^{-1}\beta))K + \overline{y}\alpha_0^\vee$ implies that $(u\Lambda_0 - \overline{y}^{-1}\beta|\alpha_0^\vee) \geq 0$. In the case of equality, $y(\dot{\alpha}_0^\vee) = \overline{y}(\alpha_0^\vee) - K \in \overline{\Delta}^\vee_+$, which contradicts (1.5.15). □

1.6. Let $\mathfrak{g}' = \mathfrak{g}(A)$ be the affine Kac–Moody algebra associated to an affine matrix A of type $X_N^{(r)}$ and let $\mathfrak{h}$ be its Cartan subalgebra. Let $\mathfrak{g}' = [\mathfrak{g}, \mathfrak{g}]$ be the derived algebra. Then $\mathfrak{h}' = \mathfrak{g}' \cap \mathfrak{h}$ is the space introduced in §1.2. The bilinear form $(.|.)$ extends from $\mathfrak{h}'$ to the whole $\mathfrak{g}(A)$ to a non–degenerate bilinear invariant symmetric form $(.|.)$ (not in a unique way). We pick an element $d \in \mathfrak{h}$, called the *scaling element*, such that

$$(d|\alpha_j) = \delta_{0j},\ j \in I,\ (d|d) = 0.$$

Then $\mathfrak{h} = \mathfrak{h}' + \mathbb{C}d$. The space $\mathfrak{h}'^*$ will be identified with a subspace of $\mathfrak{h}^*$ via extending a linear function λ on $\mathfrak{h}'$ to $\mathfrak{h}$ by $< \lambda, d >= 0$. Since the restriction of the bilinear form $(.|.)$ to $\mathfrak{h}$ is non–degenerate, it induces an isomorphism $\mathfrak{h} \tilde{\rightarrow} \mathfrak{h}^*$, and we shall identify $\mathfrak{h}$ and $\mathfrak{h}^*$ via this isomorphism. Note that d is then identified with $a_0\Lambda_0$. The action of the groups W and $\tilde{W}$ extends from $\mathfrak{h}'$ to $\mathfrak{h}$ by (1.3.1 and 2), and the bilinear form $(.|.)$ is $\tilde{W}$–invariant.

Introduce the following domains in $\mathfrak{h}$:

$$Y = \{v \in \mathfrak{h} |\ \mathrm{Re}(K|v) > 0\},$$
$$D = \{v \in \mathfrak{h} |\ \mathrm{Re}(\alpha_i|v) > 0,\ i \in I\}.$$

Note that $D \subset Y$ and recall that $\overline{D}$, the closure of D in Y in metric topology, is a fundamental domain for W in Y (see [8, Chapter 3]).

Given $\lambda \in \mathfrak{h}$, we denote by e^λ the function on $\mathfrak{h}$ defined by $e^\lambda(v) = e^{(\lambda|v)}$. The function e^{-K} will usually be denoted by q.

Define a holomorphic function R on Y by

$$R = e^\rho \prod_{\alpha \in \Delta_+} (1 - e^{-\alpha})^{\text{mult } \alpha}.$$

Here $\Delta_+ = \Delta_+^{re} \cup \mathbb{N}K$, mult $\alpha = 1$ if $\alpha \in \Delta_+^{re}$, mult $rjK = \ell$, mult $jK = (N - \ell)/(r - 1)$ if $r > 1$ and $j \not\equiv 0$ mod r (see [8, Chapters 6 and 8] for details).

Given $\Lambda \in \mathfrak{h}^*$, let $L(\Lambda)$ denote the irreducible highest weight module over $\mathfrak{g}(A)$ with highest weight Λ (see the Introduction). With respect to d we have the *energy decomposition:*

$$L(\Lambda) = \oplus_{j \in \mathbb{Z}_+} L(\Lambda)_{(\Lambda(d)-j)},$$

where $L(\Lambda)_{(a)}$ is the a–eigenspace of d in $L(\Lambda)$. We shall say that elements from $L(\Lambda)_{(\Lambda(d)-j)}$ have *energy* j. Let

$$\mathrm{ch}_\Lambda v = \mathrm{tr}_{L(\Lambda)} e^v := \sum_{\lambda \in \mathfrak{h}^*} \mathrm{mult}_\Lambda(\lambda) e^{(\lambda|v)}, \ v \in \mathfrak{h},$$

be the *character* of $L(\Lambda)$. By [8, Lemma 10.6b and (11.10.1)], this series converges in D to a holomorphic function. The following result is a special case of [12, Theorem 1]:

PROPOSITION 1.6. *Let Λ be a principal admissible weight. Then in the domain D, ch_Λ is given by the following formula:*

(1.6.1)
$$ch_\Lambda = \sum_{w \in W^\Lambda} \varepsilon(w) e^{w(\Lambda+\rho)}/R. \ \square$$

Of course, a special case of (1.6.1) when $\Lambda = 0$ is the Macdonald identity

(1.6.2)
$$\sum_{w \in W} \varepsilon(w) e^{w(\rho)} = R.$$

Remark 1.6. Both the numerator and the denominator in (1.6.1) converge to holomorphic functions in Y and the denominator R does not vanish in D.

Define the *modular anomaly* of $\Lambda \in \mathfrak{h}'^{*k}$ by

(1.6.3)
$$m_\Lambda = \frac{|\Lambda+\rho|^2}{2(k+h^\vee)} - \frac{|\rho|^2}{2h^\vee},$$

and the *normalized character* by

(1.6.4)
$$\chi_\Lambda = e^{-m_\Lambda K} \ \mathrm{ch}_\Lambda.$$

Note that χ_Λ depends only on Λ mod $\mathbb{C}K$. Another advantage of introducing the normalized character is its modular invariance properties which will be discussed later. Note that using (1.6.1) functions ch_Λ and χ_Λ can be extended to meromorphic functions in Y, which are analytic at least on the set of regular elements $Y_{\mathrm{reg}} = \{v \in Y | (\alpha|v) \neq 0 \text{ for all } \alpha \in \Delta_+^{re}\}$.

Introduce coordinates (τ, z, t) on Y as follows:

$$Y = \{(\tau, z, t) := 2\pi i(-\tau\Lambda_0 + z + tK) | \tau, t \in \mathbb{C}, \mathrm{Im}\ \tau > 0, \ z \in \overline{\mathfrak{h}}\}.$$

Then

$$q := e^{-K} = e^{2\pi i \tau}.$$

For $\lambda \in P_+^u$, $u \in \mathbf{N}$, let

$$A_\lambda = q^{|\lambda|^2/2u} \sum_{w \in W} \varepsilon(w) e^{w(\lambda)}. \tag{1.6.5}$$

This is holomorphic function on Y (which is identically zero if λ is not regular).

Let $\Lambda \in P_{u,y}^k$, $y = t_\beta \overline{y}$. Formula (1.6.1) may be rewritten in terms of the functions A_λ as follows [13, Theorem 3.5]:

$$\chi_\Lambda(\tau, z, t) = A_{\Lambda^0+\rho}(u\overline{y}^{-1} t_{-\beta/u}(\tau, z/u, t/u^2))/A_\rho(\tau, z, t). \tag{1.6.6}$$

Using this formula, it is straightforward to derive the behaviour of the characters under the involution $\Lambda \longmapsto {}^t\Lambda$:

$$\chi_{{}^t\Lambda}(\tau, z, t) = \varepsilon(\overline{w}^\Lambda)\varepsilon(\overline{w}^0)\chi_\Lambda(\tau, -z, t). \tag{1.6.7}$$

For $\mu \in \mathfrak{h}'^{*m}$, $m \neq 0$, we have the following simple but useful identity:

$$A_\lambda(m\tau, -\tau\overline{\mu}, \tau\frac{|\mu|^2}{2m}) = \sum_{w \in W} \varepsilon(w) q^{\frac{um}{2}|\frac{w(\lambda)}{u} - \frac{\mu}{m}|^2}. \tag{1.6.8}$$

Now (1.6.8) for $\mu = \rho$ and (1.6.2) give the following identity:

$$\begin{aligned} A_\lambda(h^\vee\tau, -\tau\overline{\rho}, \tau|\rho|^2/2h^\vee) &= \sum_{w \in W} \varepsilon(w) q^{|h^\vee w(\lambda) - u\rho|^2/2uh^\vee} \\ &= q^{|h^\vee\lambda - u\rho|^2/2uh^\vee} \prod_{\alpha \in \Delta_+} (1 - q^{(\lambda|\alpha)})^{\text{mult } \alpha}. \end{aligned} \tag{1.6.9}$$

Here and further, given $\lambda \in \mathfrak{h}^*$, we denote by $\overline{\lambda} \in \overline{\mathfrak{h}}$ the restriction of λ to $\overline{\mathfrak{h}}$. Finally, [13, Lemma 3.1] gives us the following asymptotics as $\tau \downarrow 0$:

$$\begin{aligned} A_\lambda(m\tau, -\tau\overline{\mu}, \tau|\mu|^2/2m) \sim (um)^{-\ell/2}|M^*/M|^{-1/2} &\prod_{\alpha \in \overline{R}_+} 4 \sin\frac{\pi(\overline{\lambda}|\alpha)}{u} \sin\frac{\pi(\overline{\mu}|\alpha)}{m} \\ &\times (-i\tau)^{-\ell/2} \exp -\frac{\pi i}{12\tau}\frac{h^\vee \dim \mathfrak{g}(X_N)}{rum}, \end{aligned} \tag{1.6.10}$$

where $\overline{R}_+$ is defined by

$$\overline{R}_+ = \overline{\Delta}_+^\vee \text{ if } r^\vee = 1, \text{ and } = \overline{\Delta}_+ \text{ if } r^\vee > 1. \tag{1.6.11}$$

§2. Branching functions for winding subalgebras.

2.1. Let $u \in \mathbb{N}$ be relatively prime to a_0, and let $y \in \tilde{W}$ satisfy (1.5.6). Denote by $\mathfrak{g}_{[u,y]}$ the subalgebra of the affine algebra $\mathfrak{g}$ generated by $\mathfrak{h}$ and the root vectors attached to the roots from $\pm y(\Pi_{[u]})$, and let $\mathfrak{g}_{[u]} = \mathfrak{g}_{[u,1]}$. The subalgebras $\mathfrak{g}_{[u,y]}$ will be called *winding subalgebras.*

Example 2.1. Let A be of type $X_\ell^{(1)}$. This is usually referredto as the non–twisted case. Let $\bar{\mathfrak{g}}$ be the simple finite–dimensional Lie algebra of type X_ℓ. Then

$$\mathfrak{g} = \mathfrak{g}(A) = \mathbb{C}[t, t^{-1}] \otimes_{\mathbb{C}} \bar{\mathfrak{g}} + \mathbb{C}K + \mathbb{C}d$$

with well–known commutation relations (see the Inroduction), and one has

$$\mathfrak{g}_{[u]} = \mathbb{C}[t^u, t^{-u}] \otimes \bar{\mathfrak{g}} + \mathbb{C}K + \mathbb{C}d.$$

One of the main objectives of the paper is to describe the decomposition of an integrable $\mathfrak{g}$–module $L(\Lambda)$ with respect to the subalgebra $\mathfrak{g}_{[u,y]}$. It turns out that this can be obtained in terms of the so called string functions. Recall that, given $\lambda \in \mathfrak{h}^*$, the *string function* c_λ^Λ of the $\mathfrak{g}$–module $L(\Lambda)$ of level k is defined by:

(2.1.1)
$$c_\lambda^\Lambda = q^{m_\Lambda - \frac{|\lambda|^2}{2k}} \sum_{n \in \mathbb{Z}} \operatorname{mult}_\Lambda(\lambda - nK) q^n.$$

(Recall that $q = e^{-K}$.) This series converges to a holomorphic function in Y [, Chapter 11]. Note that c_λ^Λ depends on $\lambda \bmod \mathbb{C}K$, that $c_\lambda^\Lambda \neq 0$ implies $\Lambda - \lambda \in Q$, and that $c_{w\lambda}^\Lambda = c_\lambda^\Lambda$ for $w \in W$.

Let $k \in \mathbb{N}$ and let $\Lambda \in P_+^k$. Then by [8, Proposition 11.8], the $\mathfrak{g}$–module $L(\Lambda)$ decomposes as a $\mathfrak{g}_{[u,y]}$–module into a direct sum of integrable irreducible highest weight modules $\dot{L}(\lambda)$ of level uk (by (1.4.3)), each appearing with a finite multiplicity. Denote by $[\Lambda : \lambda]$ the multiplicity of the occurrence of $\dot{L}(\lambda)$ in this decomposition. Introduce the *branching function* $b_\lambda^\Lambda = b_\lambda^\Lambda(\mathfrak{g}_{[u,y]})$ for the winding subalgebra, where $\Lambda \in P_+^k$, $\lambda \in y(P_+^{uk})$, by

(2.1.2)
$$b_\lambda^\Lambda = q^{m_\Lambda - u m_\lambda} \sum_{n \in \mathbb{Z}} [\Lambda : (\lambda - nK)] q^n.$$

This series converges to a holomorphic function in Y. Note that as before, b_λ^Λ depends on $\lambda \bmod \mathbb{C}K$. Note also

(2.1.3)
$$b_\lambda^\Lambda \neq 0 \text{ implies } \lambda - \Lambda \in (u-1)k\Lambda_0 + Q.$$

By the Weyl–Kac character formula (which is a special case of (1.6.1) when $\Lambda \in P_+$) we have for an integrable $\mathfrak{g}$–module $L(\Lambda)$:

(2.1.4)
$$\chi_\Lambda = A_{\Lambda+\rho}/A_\rho.$$

Similarly we have for an integrable $\mathfrak{g}_{[u,y]}$–module $\dot{L}(\lambda)$:

$$\dot{\chi}_\lambda = \dot{A}_{y(\dot{\lambda}+\dot{\rho})}/\dot{A}_{y(\dot{\rho})}, \tag{2.1.5}$$

where for $\mu \in P_+^s$, $\mu = \sum_i m_i\Lambda_i$, one lets $\dot{\mu} = \sum_i m_i\dot{\Lambda}_i$, and $\dot{\Lambda}_i$ and $\dot{\rho}$ are defined by (1.4.4 and 5), and where

$$\dot{A}_{y(\dot{\mu})} = e^{\frac{|\dot{\mu}|^2}{2s}\dot{K}} \sum_{w \in y\dot{W}y^{-1}} \varepsilon(w) e^{w(y\dot{\mu})}. \tag{2.1.6}$$

Note that

$$\dot{\mu} = \mu + (u-1)s\Lambda_0. \tag{2.1.7}$$

Finally, by the very definitions we have

$$\chi_\Lambda = \sum_{\lambda \in P_+^{uk}} b^\Lambda_{y(\lambda)} \dot{\chi}_\lambda. \tag{2.1.8}$$

We can prove now the main result of this section.

THEOREM 2.1. *Let $\mathfrak{g}$ be an affine algebra of type $X_N^{(r)}$, let $u \in \mathbb{N}$, $(u, a_0) = 1$ and let $y \in \tilde{W}$ satisfy (1.5.6). Let $\Lambda \in P_+^k$ and let $\lambda \in P_+^{uk}$. Then*

$$b^\Lambda_{y(\lambda)}(\mathfrak{g}_{[u,y]}) = \sum_{w \in W} \varepsilon(w) q^{\frac{h^\vee(uk+h^\vee)}{2k}\left|\frac{w(\lambda+\rho)}{uk+h^\vee} - \frac{\rho}{h^\vee}\right|^2} c^\Lambda_{y(w(\lambda+\rho)-\rho-(u-1)k\Lambda_0)}. \tag{2.1.9}$$

PROOF: In view of (2.1.8) and (2.1.5), it suffices to show

$$\dot{A}_{y(\dot{\rho})}\chi_\Lambda = \sum_{\lambda \in P_+^{uk}} b^\Lambda_{y(\lambda)} \dot{A}_{y(\dot{\lambda}+\dot{\rho})}, \tag{2.1.10}$$

where $b^\Lambda_{y(\lambda)}$ is given by the right–hand side of (2.1.9).

Recall that by definition of string functions we have [8, Chapter 12]:

$$\chi_\Lambda = \sum_{\substack{\xi \in \Lambda+Q+\mathbb{C}K \\ \mathrm{mod}\ (kM+\mathbb{C}K)}} \sum_{\gamma \in M} q^{\frac{|\xi|^2}{2k}} e^{t_\gamma(\xi)} c^\Lambda_\xi. \tag{2.1.11}$$

The left–hand side of (2.1.10) is equal to

$$e^{-\frac{|\dot{\rho}|^2}{2h^\vee}\dot{K}} \sum_{w \in y\dot{W}y^{-1}} \varepsilon(w) e^{w(y\dot{\rho})} \chi_\Lambda. \tag{2.1.12}$$

We can write $w = yt_{u\alpha}\overline{w}y^{-1}$, where $\alpha \in M$, $\overline{w} \in \overline{W}$. Letting $\gamma = \overline{y}\overline{w}\gamma'$, $\xi = w^{-1}y^{-1}\xi'$ and using Lemma 1.4 and (1.4.3–6), we can rewrite (2.1.12) as follows

(2.1.13)
$$\begin{aligned} &q^{\frac{u|\rho|^2}{2h^\vee}} \sum_{w\in y\dot{W}y^{-1}} \sum_{\substack{\xi'\in(wy)^{-1}\Lambda+Q+\mathbf{C}K \\ \mathrm{mod}\ (kM+\mathbf{C}K)}} \sum_{\gamma\in M} \varepsilon(w) q^{\frac{|\xi'|^2}{2k}} e^{wy(\dot{\rho}+t_\gamma\xi')} c^{\Lambda}_{wy\xi'} \\ &= q^{\frac{u|\rho|^2}{2h^\vee}} \sum_{w\in y\dot{W}y^{-1}} \sum_{\substack{\xi\in y^{-1}\Lambda+Q+\mathbf{C}K \\ \mathrm{mod}\ (kM+\mathbf{C}K)}} \sum_{\gamma\in M} \varepsilon(w) q^{\frac{|\xi+k\gamma|^2}{2k}} e^{wy(\dot{\rho}+\xi+k\gamma)} c^{\Lambda}_{y\xi}. \end{aligned}$$

We may assume that $y(\dot{\rho}+\xi+k\gamma)$ is regular with respect to $y\dot{W}y^{-1}$. Then there exists a unique element $\tilde{\nu}$ from the set $y(P_+^{uk})$ of dominant integral weights of level uk for the basis $y\Pi^\vee_{[u]}y^{-1}$ and a unique $\sigma \in y\dot{W}y^{-1}$ such that

(2.1.14)
$$y(\dot{\rho}+\xi+k\gamma) = \sigma(\tilde{\nu}+y(\dot{\rho})) + aK, \text{ where } a \in \mathbf{C}.$$

Plugging (2.1.14) in (2.1.13) we obtain:

(2.1.15)
$$\sum_{w\in y\dot{W}y^{-1}} \sum_{\sigma\in y\dot{W}y^{-1}} \sum_{\tilde{\nu}\in y(P_+^{uk})} \varepsilon(w) e^{w\sigma(\tilde{\nu}+y\dot{\rho})} q^b c^{\Lambda}_{\sigma(\tilde{\nu}+y\dot{\rho})-y\dot{\rho}}$$

where

$$\begin{aligned} b &= -a + \frac{|\sigma(\tilde{\nu}+y\dot{\rho}) - y\dot{\rho} + aK|^2}{2k} + \frac{u|\rho|^2}{2h^\vee} \\ &= \frac{|\sigma(\tilde{\nu}+y\dot{\rho}) - y\dot{\rho}|^2}{2k} + \frac{u|\rho|^2}{2h^\vee} \\ &= \frac{h^\vee(uk+h^\vee)}{2k}\left|\frac{\sigma(\tilde{\nu}+y\dot{\rho})}{uk+h^\vee} - \frac{y(\dot{\rho})}{h^\vee}\right|^2 + \frac{u|\tilde{\nu}+y\dot{\rho}|^2}{2(uk+h^\vee)}. \end{aligned}$$

Plugging this expression of b in (2.1.15) and putting $w' = w\sigma$, $\dot{\nu} = y^{-1}\tilde{\nu}$, we obtain:

(2.1.16)
$$\sum_{\dot{\nu}\in P_+^{uk}} \sum_{\sigma\in y\dot{W}y^{-1}} \varepsilon(\sigma)\dot{A}_{y(\dot{\nu}+\dot{\rho})} q^{\frac{h^\vee(uk+h^\vee)}{2k}\left|\frac{\sigma y(\dot{\nu}+\dot{\rho})}{uk+h^\vee} - \frac{y(\dot{\rho})}{h^\vee}\right|^2} c^{\Lambda}_{\sigma y(\dot{\nu}+\dot{\rho})-y\dot{\rho}}.$$

Using (2.1.7) completes the proof. $\square$

Since $c^\Lambda_\lambda = c^\Lambda_{w(\lambda)}$ for $w \in W$, we obtain, in view of (1.3.4), the following corollary of (2.1.9):

COROLLARY 2.1. *Decompose $y \in \tilde{W}$ according to (1.3.4): $y = \tilde{y}w$, $\tilde{y} \in \tilde{W}_+$, $w \in W$. Then*

$$b_\lambda^\Lambda(\mathfrak{g}_{[u,y]}) = b_\lambda^{\tilde{y}^{-1}(\Lambda)}(\mathfrak{g}_{[u]}). \quad \square$$

This corollary allows one to reduce the study of branching functions for the subalgebra $\mathfrak{g}_{[u,y]}$ to that for the subalgebra $\mathfrak{g}_{[u]}$.

Since $b_\lambda^\Lambda(\mathfrak{g}_{[1]}) = \delta_{\Lambda,\lambda}$, we obtain another corollary of (2.1.9). For $\Lambda, \lambda \in P_+^k$ one has:

$$\sum_{w \in W} \varepsilon(w) q^{\frac{h^\vee(k+h^\vee)}{2k}\left|\frac{w(\lambda+\rho)}{k+h^\vee} - \frac{\rho}{h^\vee}\right|^2} c^\Lambda_{w(\lambda+\rho)-\rho} = \delta_{\Lambda,\lambda}. \tag{2.1.17}$$

2.2. In this subsection we consider the special case $k = 1$ of Theorem 2.1. Recall the following

LEMMA 2.2. *([10, 4.6], [8, Chapter 13]). Let $\mathfrak{g}$ be an affine algebra of type $X_N^{(r)}$ where X_N is of type A_N, D_N or E_N, and let $\Lambda, M \in P_+^1 + \mathbb{C}K$. Then either $\Lambda - M \notin Q$ and $c_M^\Lambda = 0$, or $\Lambda - M \in Q$ and*

$$c_M^\Lambda = q^{m_\Lambda - |\Lambda|^2/2} \prod_{n \in \mathbb{N}} (1 - q^n)^{-\text{mult } nK}. \quad \square$$

Using Lemma 2.2 and (1.6.9), formula (2.1.9) can be rewritten in the case $\Lambda \in P_+^1$ as follows:

THEOREM 2.2. *Let $\mathfrak{g}$ be an affine algebra of type $X_N^{(r)}$, where X_N is of type A_N, D_N or E_N (i.e. either $r = r^\vee = 1$ or $r > 1$), and let $\Lambda \in P_+^1$. Let $u \in \mathbb{N}$, let $\lambda \in P_+^u$, and consider the subalgebra $\mathfrak{g}_{[u]}$ of $\mathfrak{g}$. Then either $\Lambda - (\lambda - (u-1)\Lambda_0) \notin Q$ and $[\Lambda : (\lambda - nK)] = 0$ for all n, or $\Lambda - (\lambda - (u-1)\Lambda_0) \in Q$ and*

$$\sum_{n \in \mathbb{Z}} [\Lambda : (\lambda - nK)] q^n = \frac{q^{(|\lambda|^2 - |\Lambda|^2)/2} \prod_{\alpha \in \Delta_+} (1 - q^{(\lambda+\rho|\alpha)})^{\text{mult } \alpha}}{\prod_{n \in \mathbb{N}} (1 - q^n)^{\text{mult } nK}}. \quad \square \tag{2.2.1}$$

This result was proved in the case of $A_\ell^{(1)}$ in [4] by a quite complicated method and was conjectured there for the simply laced case (except that the power of q is missing there).

COROLLARY 2.2. *Let $\mathfrak{g}$ be as in Theorem 2.2 and let $u \in \mathbb{N}$.*

(a) If $\Lambda \in P_+^1$ and $\lambda \in P_+^u$ are such that $\lambda - \Lambda \in (u-1)\Lambda_0 + Q$ then the minimal n for which $[\Lambda : (\lambda - nK)] \neq 0$ is equal to $\frac{1}{2}(|\lambda|^2 - |\Lambda|^2)$, and $[\Lambda : (\lambda - \frac{1}{2}(|\lambda|^2 - |\Lambda|^2)K)] = 1$.

(b) In the $\mathfrak{g}$-module $\oplus_{\Lambda \in P_+^1} L(\Lambda)$ viewed as a $\mathfrak{g}'_{[u]}$-module all integrable highest weight modules of level u occur, the minimal energy of occurring of the highest weight vector of $L(\lambda)$, $\lambda \in P_+^u$, being equal to $\frac{1}{2}(|\overline{\lambda}|^2 - |\overline{\Lambda}|^2)$. $\square$

According to Corollary 2.2b, given $\mathfrak{g}$ as in Theorem 2.2 and $u \in \mathbb{N}$, for each $\lambda \in P_+^u$ there exists a canonically defined 1-dimensional subspace in the space $V = \oplus_{\Lambda \in P_+^1} L(\Lambda)$, which we denote by $V_\lambda^{[u]}$, whose non-zero vectors are all highest weight vectors of weight λ for $\mathfrak{g}'_{[u]}$ of lowest energy. (Comparing this with [8, Exercise 12.17], we see that $V_\lambda^{[u]}\backslash\{0\}$ is also the set in V of highest weight vectors of weight $\overline{\lambda}$ for $\mathring{\mathfrak{g}}$ of lowest energy.) Since all weights of $L(\Lambda)$, $\Lambda \in P_+^1$, are of the form $t_\gamma(\Lambda) - nK$, $\gamma \in \overline{Q}$, $n \in \mathbb{Z}_+$, we conclude that

$$V_\lambda^{[u]} = L(\Lambda)_{t_\gamma(\Lambda)}, \text{ where } \lambda - \Lambda = (u-1)\Lambda_0 + \gamma \bmod \mathbb{C}K. \tag{2.2.2}$$

Let $N' = N$ for $A = X_N^{(r)}$ except for $A = A_{2\ell-1}^{(2)}$ and $D_{\ell+1}^{(2)}$, when we let $N' = \ell + 1$ and $2\ell - 1$ respectively. Introduce the following function,

$$G(\tau) = q^{N'/24a_0} \prod_{n\geq 1} (1-q^n)^{\text{mult } nK}.$$

This function, along with the related values of $|\overline{Q}^\vee/M|$, is given by the following table:

$Type X_N^{(r)}$	$G(\tau)$	$\lvert\overline{Q}^\vee/M\rvert$
$X_\ell^{(1)}$ or $A_{2\ell}^{(2)}$	$\eta(\tau)^\ell$	1
$A_{2\ell-1}^{(2)}$	$\eta(\tau)^{\ell-1}\eta(2\tau)$	2
$D_{\ell+1}^{(2)}$	$\eta(\tau)\eta(2\tau)^{\ell-1}$	$2^{\ell-1}$
$E_6^{(2)}$	$\eta(\tau)^2\eta(2\tau)^2$	4
$D_4^{(3)}$	$\eta(\tau)\eta(3\tau)$	3

Then formula (2.2.1) can be rewritten as follows:

$$b_\lambda^\Lambda = \frac{q^{|h^\vee\lambda - u\rho|^2/2(u+h^\vee)h^\vee} \prod_{\alpha\in\Delta_+}(1-q^{(\lambda+\rho|\alpha)})^{\text{mult }\alpha}}{G(\tau)}. \tag{2.2.3}$$

Remark 2.2. Formulas (1.2.3), (1.2.4), (1.3.5) and (1.3.6) together with the above table allow one to compute the values of $|M^*/M|$, $|\tilde{M}/M|$, etc.

Later we shall need the following asymptotics as $\tau \downarrow 0$:

$$G(\tau)^{-1} \sim |\overline{Q}^\vee/M|^{1/2}(-i\tau)^{\ell/2} \exp\frac{\pi i N}{12r\tau}, \tag{2.2.4}$$

which can be deduced either by using the above table and the asymptotics of $\eta(\tau)^{-1}$, or by using the asymptotics of the string function $c_{\Lambda_0}^{\Lambda_0}$ [8, Chapter 13].

There is one more case when branching functions for winding subalgebras have very simple expressions: $\mathfrak{g}$ is of type $B_\ell^{(1)}$ and $k = 1$. In this case all (up to equivalence and symmetry of the Dynkin diagram) non-zero string functions are given by the following formulas [10, 4.6]:

$$c_{\Lambda_0}^{\Lambda_0} + c_{\Lambda_1}^{\Lambda_0} = q^{m_{\Lambda_0}} \prod_{n\in\mathbb{N}} (1-q^n)^{-\ell}(1+q^{n-1/2}), \tag{2.2.5}$$

(2.2.6) $$c_{\Lambda_\ell}^{\Lambda_\ell} = q^{m_{\Lambda_\ell} - |\Lambda_\ell|^2/2} \prod_{n\in\mathbb{N}} (1-q^n)^{-\ell}(1+q^n).$$

THEOREM 2.2. *' Let $\mathfrak{g}$ be an affine algebra of type $B_\ell^{(1)}$ and let $\Lambda \in P_+^1 = \{\Lambda_0, \Lambda_1, \Lambda_\ell\}$. Let $u \in \mathbb{N}$, $\lambda \in P_+^u$ and consider the subalgebra $\mathfrak{g}_{[u]}$ of $\mathfrak{g}$. Then either all branching coefficients $[\Lambda : (\lambda - nK)]$ are zero, or $\lambda \in u\Lambda_0 + Q$ (resp. $\in \Lambda_\ell + (u-1)\Lambda_0 + Q$) and we have respectively:*

(2.2.7) $$\begin{aligned}&\sum_{n\in\mathbb{Z}}[\Lambda_0 :\ (\lambda - nK)]q^n + \sum_{n\in\mathbb{Z}}[\Lambda_1 :\ (\lambda - nK)]q^{n+1/2} \\ &= q^{|\lambda|^2/2} \prod_{\alpha\in\Delta_+} (1 - q^{(\lambda+\rho|\alpha)})^{\text{mult}\,\alpha} \prod_{n\in\mathbb{N}} (1-q^n)^{-\ell}(1+q^{n-1/2}),\end{aligned}$$

(2.2.8)
$$\begin{aligned}&\sum_{n\in\mathbb{Z}}[\Lambda_\ell :\ (\lambda - nK)]q^n \\ &\qquad = q^{(|\lambda|^2 - |\Lambda_\ell|^2)/2} \prod_{\alpha\in\Delta_+} (1 - q^{(\lambda+\rho|\alpha)})^{\text{mult}\,\alpha} \prod_{n\in\mathbb{N}} (1-q^n)^{-\ell}(1+q^n).\end{aligned}$$

PROOF: Formula (2.2.8) follows from (2.1.9), (1.6.9), (2.2.6) and the fact that any non–zero string function of the form $c_\lambda^{\Lambda_\ell}$ is equal to $c_{\Lambda_\ell}^{\Lambda_\ell}$.

Adding up formulas (2.1.9) for $\Lambda = \Lambda_0$ and Λ_1, we have:

(2.2.9)
$$b_\lambda^{\Lambda_0} + b_\lambda^{\Lambda_1} = \sum_{w\in W} \varepsilon(w) q^{\frac{h^\vee(u+h^\vee)}{2}|\frac{w(\lambda+\rho)}{u+h^\vee} - \frac{\rho}{h^\vee}|^2} (c^{\Lambda_0}_{w(\lambda+\rho)-\rho-(u-1)\Lambda_0} + c^{\Lambda_1}_{w(\lambda+\rho)-\rho-(u-1)\Lambda_0}).$$

Since $w(\lambda+\rho) - \rho - (u-1)\Lambda_0 \in P^1$, it is Q–congruent to Λ_0, hence is W–conjugate to Λ_0 or Λ_1. Hence $c^{\Lambda_0}_{w(\lambda+\rho)-\rho-(u-1)\Lambda_0} + c^{\Lambda_1}_{w(\lambda+\rho)-\rho-(u-1)\Lambda_0} = c^{\Lambda_0}_{\Lambda_0} + c^{\Lambda_1}_{\Lambda_0}$ or $= c^{\Lambda_0}_{\Lambda_1} + c^{\Lambda_1}_{\Lambda_1}$, which are equal. Now we can apply to (2.2.9) formulas (1.6.9) and (2.2.5). □

2.3. In the section we discuss the asymptotics of the branching functions $b_\lambda^\Lambda(\mathfrak{g}_{[u]};\tau)$ as $\tau \downarrow 0$.

Introduce the following notation (see (1.6.11)):

(2.3.1) $$a(\Lambda) = (k+h^\vee)^{-\ell/2} |M^*/\overline{Q}^\vee|^{-\frac{1}{2}} \prod_{\alpha\in\overline{R}_+} 2\sin\frac{\pi(\Lambda+\rho|\alpha)}{k+h^\vee},\ \Lambda \in P_+^k;$$

(2.3.2) $$c_k = \frac{k \dim \mathfrak{g}(X_N)}{k+h^\vee},\ k\in\mathbb{Q}.$$

We can state now the following

CONJECTURE 2.2. *Let $\mathfrak{g}$ be an affine algebra of type $X_N^{(r)}$, let $k \in \mathbb{N}$, $\Lambda \in P_+^k$ and let $u \in \mathbb{N}$, $u > 1$, $\lambda \in P_+^{uk}$ be such that $\lambda - \Lambda \in (u-1)k\Lambda_0 + Q$. Then*

$$b_\lambda^\Lambda(\mathfrak{g}_{[u]};\tau) \sim |J|a(\Lambda)a(\lambda)\exp\frac{\pi i}{12\tau r}(c_k - u^{-1}c_{uk}). \tag{2.3.3}$$

Note that asymptotics (2.3.3) is obtained from (2.1.9) if one replaces thoughtlessly the string functions by their asymptotics (given in [10, §4.7] or [8, Chapter 13]):

$$c_\lambda^\Lambda(\tau) \sim |J||\overline{Q}^{\vee *}/M|^{-1}|M^*/M|^{1/2}a(\Lambda)(-i\tau/k)^{\ell/2}e^{\pi i c_k/12r\tau} \text{ if } \Lambda - \lambda \in Q, \tag{2.3.4}$$

and then uses (1.6.8 and 10). This procedure gives however a wrong result when $u = 1$. A more careful argument using the second term in the asymptotic expansion of the c_λ^Λ shows that (2.3.3) holds for sufficiently large u. More precisely, let

$$b_k = \min_{\lambda \in P_+^k \setminus \{k\Lambda_j, j \in J\}} (2k(\lambda|\rho) - h^\vee|\lambda|^2)k^2,$$

and let, as usual, b'_k denote the same quantity for the adjacent root system ([10,§1.5] or [13, §3]). (Note that $b_k = b'_k$ if $r = 1$ or $a_0 = 2$. Recall also that $b_k > 0$ [10, Proposition 4.14]. For example, $b_k = \ell(k-1)/k^2$ for $A_\ell^{(1)}$, $k > 1$.) Then (2.3.3) holds if

$$12a_0b'_k/(k+h^\vee) > \dim \mathfrak{g}(X_N)/(uk+h^\vee).$$

Note also that (2.3.3) holds for $k = 1$ if $r = r^\vee = 1$ or $r > 1$. This follows from Lemma 2.2 by making use of the "thoughtless" argument.

2.4. In this section we will assume (for simplicity) that $r = 1$, hence $\mathfrak{g}' = \mathbb{C}[t,t^{-1}] \otimes_{\mathbb{C}} \mathring{\mathfrak{g}} + \mathbb{C}K$ as described in Example 2.1. The well-known Sugawara construction extends any representation of $\mathfrak{g}$ in a vector space V with $K = kI_V$ and with spectrum of d bounded below, to the semidirect product Vir $\ltimes \mathfrak{g}'$ (see e.g. [11, §3.4] where this is described also in the twisted case). Recall that Vir is spanned by operators L_n ($n \in \mathbb{Z}$) and I, which satisfy the usual Virasoro relations:

$$[L_m, L_n] = (m-n)L_{m+n} + \delta_{m,-n}\frac{m^3-m}{12}c_k,$$

where the *central charge* c_k is given by (2.3.2), and that $[t^m \otimes x, L_n] = mt^{m+n} \otimes x$.

We fix $u \in \mathbb{N}$ and let

$$\tilde{L}_n = u^{-1}L_{un} + \delta_{n,0}(u - u^{-1})c_k/24.$$

These operators satisfy the Virasoro relations with central charge uc_k, and for $x(um) \in \mathfrak{g}'_{[u]}$ we have: $[x(um), \tilde{L}_n] = mx(u(m+n)) \in \mathfrak{g}'_{[u]}$. (Recall that $x(m)$ stands for $t^m \otimes x$.)

Let $\dot{L}_n$ ($n \in \mathbb{Z}$) be the operators given by the Sugawara construction for the $\mathfrak{g}'_{[u]}$–module V. They satisfy Virasoro relations with central charge c_{uk}. Hence the coset Virasoro operators

$$L_n^{[u]} := \tilde{L}_n - \dot{L}_n \tag{2.4.1}$$

satisfy the Virasoro relations with central charge

$$c_k^{[u]} := uc_k - c_{uk}. \tag{2.4.2}$$

This is a variation of the well–known coset construction [5].

The operators $L_n^{[u]}$ commute with $\mathfrak{g}'_{[u]}$, hence act on each subspace $\mathcal{U}(\Lambda, \lambda)$ of highest weight vectors with weight $\lambda \in P_+^{ku}$ for $\mathfrak{g}'_{[u]}$ in a $\mathfrak{g}$–module $L(\Lambda)$, $\Lambda \in P_+^k$. This representation of Vir in $\mathcal{U}(\Lambda, \lambda)$ is unitary with central charge $c_k^{[u]}$ and with the lowest eigenvalue of $L_0^{[u]}$ equal to

$$h_{\Lambda,\lambda}^{[u]} = u^{-1} m_\Lambda - \dot{m}_\lambda + c_k^{[u]}/24 + u^{-1} n, \tag{2.4.3}$$

where n is the lowest energy of d in $\mathcal{U}(\Lambda, \lambda)$.

Using the above facts it is easy to compute the branching functions in terms of Virasoro characters $\chi_{r,s}^{(m)}$ in the case when $c_k^{[u]} < 1$, using the method explained in [11] (we use notation of [11]).

PROPOSITION 2.4. *Let $\mathfrak{g} = \mathfrak{g}(A)$ be an affine algebra associated to a symmetric matrix A and let $u > 1$. Let $c = c_k^{[u]} < 1$; then $k = 1$ and $u = 2$, and all these cases are listed below (in all formulas $b = b(\tau)$ and $\chi = \chi(2\tau)$):*

0) $A_1^{(1)}$, $c = \frac{1}{2}$:
$b_{2\Lambda_0}^{\Lambda_0} = b_{2\Lambda_1}^{\Lambda_0} = \chi_{2,2}^{(1)}$; $b_{\Lambda_0+\Lambda_1}^{\Lambda_1} = \chi_{1,1}^{(1)} + \chi_{2,1}^{(1)}$.
1) $E_8^{(1)}$, $c = \frac{1}{2}$:
$b_{2\Lambda_0}^{\Lambda_0} = \chi_{2,1}^{(1)}$; $b_{\Lambda_1}^{\Lambda_0} = \chi_{2,2}^{(1)}$; $b_{\Lambda_7}^{\Lambda_0} = \chi_{1,1}^{(1)}$.
2) $E_7^{(1)}$, $c = \frac{7}{10}$:
$b_{2\Lambda_0}^{\Lambda_0} = b_{2\Lambda_6}^{\Lambda_0} = \chi_{2,1}^{(2)}$; $b_{\Lambda_1}^{\Lambda_0} = b_{\Lambda_5}^{\Lambda_0} = \chi_{2,2}^{(2)}$; $b_{\Lambda_0+\Lambda_6}^{\Lambda_6} = \chi_{3,2}^{(2)} + \chi_{3,3}^{(2)}$; $b_{\Lambda_7}^{\Lambda_6} = \chi_{1,1}^{(2)} + \chi_{3,1}^{(2)}$.
3) $A_2^{(1)}$, $c = \frac{4}{5}$:
$b_{2\Lambda_0}^{\Lambda_0} = b_{2\Lambda_2}^{\Lambda_1} = b_{2\Lambda_1}^{\Lambda_2} = \chi_{4,2}^{(3)} + \chi_{4,4}^{(3)}$; $b_{\Lambda_1+\Lambda_2}^{\Lambda_0} = b_{\Lambda_0+\Lambda_1}^{\Lambda_1} = b_{\Lambda_0+\Lambda_2}^{\Lambda_2} = \chi_{2,2}^{(3)} + \chi_{3,2}^{(3)}$.
4) $E_6^{(1)}$, $c = \frac{6}{7}$:
$b_{2\Lambda_0}^{\Lambda_0} = b_{2\Lambda_5}^{\Lambda_1} = b_{2\Lambda_1}^{\Lambda_5} = \chi_{2,1}^{(4)} + \chi_{4,1}^{(4)}$, $b_{\Lambda_6}^{\Lambda_0} = b_{\Lambda_4}^{\Lambda_1} = b_{\Lambda_2}^{\Lambda_5} = \chi_{2,2}^{(4)} + \chi_{4,2}^{(4)}$,
$b_{\Lambda_1+\Lambda_5}^{\Lambda_0} = b_{\Lambda_0+\Lambda_1}^{\Lambda_1} = b_{\Lambda_0+\Lambda_5}^{\Lambda_5} = \chi_{4,3}^{(4)}$. □

Remark 2.4. Comparing (for A of type $E_8^{(1)}$, $\Lambda = \Lambda_0$, $\lambda = \Lambda_1$) formula (2.2.1) with Proposition 2.4, we obtain the following curious identity:

$$\frac{\prod_{\alpha \in \Delta_+(E_8^{(1)})} (1 - q^{(\Lambda_1 + \rho|\alpha)})^{\text{mult } \alpha}}{\prod_{n \in \mathbb{N}} (1 - q^n)^8} = \prod_{n \in \mathbb{N}} (1 + q^{2n}).$$

§3. Comparison with branching functions for tensor products.

3.1. Let $\mathfrak{g} = \mathfrak{g}(A)$ be an affine algebra of type $X_N^{(r)}$. Let $m \in \mathbb{N}$, let $k \in \mathbb{Q}$ be principal admissible, and let $\Lambda \in P_+^m$, $\mu \in P^k(A)$ (see §1.6). Then by [13, Corollary 4.1] the module $L(\Lambda) \otimes L(\mu)$ decomposes with respect to $\mathfrak{g}$ into a direct sum of irreducible highest weight modules $L(\lambda)$, where $\lambda \in P^{m+k}(A) + \mathbb{C}K$, each appearing with finite multiplicity. Denote by $[\Lambda \otimes \mu : \lambda]$ this multiplicity. Define the branching function $b_\lambda^{\Lambda\otimes\mu}$ as follows:

$$b_\lambda^{\Lambda\otimes\mu} = q^{m_\Lambda + m_\mu - m_\lambda} \sum_{n\in\mathbb{Z}} [\Lambda \otimes \mu : (\lambda - nK)] q^n.$$

Note that $b_\lambda^{\Lambda\otimes\mu} = 0$ unless $\Lambda + \mu - \lambda \in Q$.

We have seen already in §2.1 that the branching functions for winding subalgebras can be expressed via string functions. We will show now that this holds for tensor products as well. Recall the map $\lambda \longmapsto \lambda^0$ defined by (1.5.7).

THEOREM 3.1. *Let $\mathfrak{g}$ be an affine algebra of type $X_N^{(r)}$, let $m \in \mathbb{N}$ and $\Lambda \in P_+^m$, and let k be a principal admissible rational number with denominator u and $\mu = y.(\mu^0 - (u-1)(k+h^\vee)\Lambda_0) \in P_{u,y}^k$. Then $b_\lambda^{\Lambda\otimes\mu} \neq 0$ implies that $\lambda \in P_{u,y}^{m+k}$ and in this case one has:*

$$b_\lambda^{\Lambda\otimes\mu} = \sum_{w\in W} \varepsilon(w) q^{\frac{(k+h^\vee)(m+k+h^\vee)}{2m} \left| \frac{w(\lambda^0+\rho)}{m+k+h^\vee} - \frac{\mu^0+\rho}{k+h^\vee} \right|^2} c^\Lambda_{y(w(\lambda^0+\rho)-(\mu^0+\rho)-(u-1)m\Lambda_0)}. \tag{3.1.1}$$

PROOF: By the very definitions we have:

$$\chi_\Lambda \chi_\mu = \sum_{\lambda \in P_{u,y}^{m+k}} b_\lambda^{\Lambda\otimes\mu} \chi_\lambda. \tag{3.1.2}$$

By (1.6.1 and 2) we have

$$\chi_\mu = \hat{A}_{\mu+\rho} / A_\rho, \tag{3.1.3}$$

where

$$\hat{A}_{\mu+\rho} = q^{\frac{|\mu^0+\rho|^2}{2(k+h^\vee)}} \sum_{w \in y\mathring{W}y^{-1}} \varepsilon(w) e^{w(\mu+\rho)}. \tag{3.1.4}$$

A similar formula holds for χ_λ. Using (3.1.2 and 3), it suffices to show

$$\chi_\Lambda \hat{A}_{\mu+\rho} = \sum_{\lambda \in P_{u,y}^{m+k}} b_\lambda^{\Lambda\otimes\mu} \hat{A}_{\lambda+\rho}, \tag{3.1.5}$$

where $b_\lambda^{\Lambda\otimes\mu}$ is given by the right–hand side of (3.1.1). Using (2.1.11) and (3.1.4) we can rewrite the left–hand side of (3.1.5) as follows:

$$(3.1.6)\qquad q^{\frac{|\mu^0+\rho|^2}{2(k+h^\vee)}} \sum_{\substack{\xi\in\Lambda+Q+\mathbb{C}K \\ \bmod (mM+\mathbb{C}K)}} \sum_{w\in y\dot{W}y^{-1}} \varepsilon(w)e^{w(\mu+\rho)} \sum_{\gamma\in M} q^{\frac{|\xi|^2}{2m}} e^{t_\gamma(\xi)} c_\xi^\Lambda.$$

Recall that $w = yt_{u\alpha}\overline{w}y^{-1}$; let $\gamma = \overline{y}\ \overline{w}\gamma'$ and $\xi = \overline{w}\ \overline{y}\xi'$. Then, using Lemma 1.4, we can rewrite (3.1.6) as follows:

$$(3.1.7)\qquad q^{\frac{|\mu^0+\rho|^2}{2(k+h^\vee)}} \sum_{w\in y\dot{W}y^{-1}} \sum_{\substack{\xi'\in y^{-1}(\Lambda)+Q+\mathbb{C}K \\ \bmod (mM+\mathbb{C}K)}} \sum_{\gamma'\in M} \varepsilon(w)e^{w(\mu+\rho+y(\xi'+m\gamma'))} q^{\frac{|\xi'+m\gamma|^2}{2m}} c_{y(\xi')}^\Lambda.$$

We may assume that $\mu+\rho+y(\xi'+m\gamma')$ is regular with respect to $y\dot{W}y^{-1}$. Then there exists a unique dominant integral weight $\tilde{\nu}$ of level $u(m+k+h^\vee)-h^\vee$ for the simple coroot basis $y(\Pi_{[u]})$ and a unique $\sigma\in y\dot{W}y^{-1}$ such that

$$(3.1.8)\qquad (\mu+\rho+y(\xi'+m\gamma'))-\sigma(\tilde{\nu}+y(\dot{\rho}))\in\mathbb{C}K,$$

where $\dot{\rho}$ is given by (1.4.5).

We can write $y^{-1}(\tilde{\nu})$ in the form:

$$y^{-1}(\tilde{\nu}) = \sum_i m_i\dot{\Lambda}_i,\ m_i\in\mathbb{Z}_+,\ \sum_i a_i^\vee m_i = u(m+k+h^\vee)-h^\vee,$$

where the $\dot{\Lambda}_i$ are defined by (1.4.4). Let $\nu+\rho=\tilde{\nu}+y(\dot{\rho})$. Then we have by (1.4.4 and 5):

$$(3.1.9)\qquad \nu+\rho = y(\nu^0+\rho-(u-1)(m+k+h^\vee)\Lambda_0),$$

where $\nu^0 = \Sigma m_i\Lambda_i$, hence $\nu\in P_{u,y}^{m+k}$.

Using (3.1.9) we can rewrite (3.1.8) as follows:

$$(3.1.10)\qquad \mu+\rho+y(\xi'+m\gamma') = \sigma(\nu+\rho)+aK,\ a\in\mathbb{C}.$$

Since the string functions are invariant under translations by elements from mM, we see from (3.1.10), that

$$(3.1.11)\qquad c_{y(\xi')}^\Lambda = c_{\sigma(\nu+\rho)-(\mu+\rho)}^\Lambda.$$

Plugging (3.1.10) and (3.1.11) in (3.1.7) we obtain

$$(3.1.12)\qquad \sum_{\nu\in P_{u,y}^{m+k}} \sum_{\sigma\in y\dot{W}y^{-1}} \varepsilon(\sigma)\hat{A}_{\nu+\rho}q^b c_{\sigma(\nu+\rho)-(\mu+\rho)}^\Lambda,$$

where

$$\begin{aligned} b &= -a+\frac{|\sigma(\nu+\rho)-(\mu+\rho)+aK|^2}{2m}+\frac{|\mu^0+\rho|^2}{2(k+h^\vee)}-\frac{|\nu^0+\rho|^2}{2(m+k+h^\vee)} \\ &= \frac{|\sigma(\nu+\rho)-(\mu+\rho)|^2}{2m}+\frac{|\mu^0+\rho|^2}{2(k+h^\vee)}-\frac{|\nu^0+\rho|^2}{2(m+k+h^\vee)}. \end{aligned}$$

Furthermore, we have: $\sigma = yt_{u\alpha}\overline{\sigma}y^{-1}$, $\alpha\in M$, $\overline{\sigma}\in\overline{W}$. Hence, using (3.1.9), we obtain:

$$(3.1.13)\qquad \sigma(\nu+\rho) = y(t_\alpha\overline{\sigma}(\nu^0+\rho)-(u-1)(m+k+h^\vee)\Lambda_0) \bmod \mathbb{C}K.$$

Formulas (3.1.12 and 13) prove (3.1.5) and the theorem. □

CONJECTURE 3.1. *Let $\mathfrak{g}$ be an affine algebra of type $X_N^{(r)}$, let $m \in \mathbb{N}$ and $\Lambda \in P_+^m$, and let $\mu \in P_{u,y}^k$ and $\lambda \in P_{u,y}^{m+k}$ be such that $\Lambda + \mu - \lambda \in Q$. Then*

$$b_\lambda^{\Lambda\otimes\mu}(\tau) \sim |J| a(\Lambda) a(\lambda^0) a(\mu^0) \exp \frac{\pi i}{12\tau r}(c_m + c_k - c_{m+k}). \tag{3.1.14}$$

Note that asymptotics (3.1.14) is proved in the case $k \in \mathbb{N}$ [11, (2.7.15)]. As in the case of Conjecture 2.1, one can show that (3.1.14) holds for sufficiently large u, namely, when the following inequality is satisfied:

$$\frac{12 a_0 b'_m}{m + h^\vee} > \frac{h^\vee \dim \mathfrak{g}(X_N)}{u^2(k+h^\vee)(m+k+h^\vee)}.$$

3.2. We shall compare now Theorems 2.1 and 3.1.

PROPOSITION 3.2. *Let $\mathfrak{g}, m$ and Λ be as in Theorem 3.1. Let $u \in \mathbb{N}$ be such that $(u, h^\vee) = 1$ and $(u, r^\vee) = 1$. Let $k = (u^{-1} - 1)h^\vee$ and let $y \in \tilde{W}$ satisfy (1.5.6). Then (by Proposition 1.5) $P_{u,y}^k = \{\mu := y.(k\Lambda_0)\}$, and for any $\lambda = y.(\lambda^0 - (u-1)(m + u^{-1}h^\vee)\Lambda_0) \in P_{u,y}^{m+k}$ (where $\lambda^0 \in P_+^{um}$) one has:*

$$b_\lambda^{\Lambda\otimes\mu} = b_{\lambda^0}^{\Lambda}(\mathfrak{g}_{[u,y]}). \tag{3.2.1}$$

PROOF: Just compare formulas (3.1.1) and (2.1.9). □

§4. Functions $\varphi_{\lambda,\mu}$ and modular invariance.

4.1. Branching functions for winding subalgebras are intimately related to the functions $\varphi_{\lambda,\mu}$ defined as follows. As in Lemma 1.5, we shall distinguish cases 1) $r = r^\vee = 1$ (i.e. A is symmetric) or $r > 1$, and 2) $r = 1$ but $r^\vee > 1$, putting the second case in parenthesis. Let p and p' be positive integers such that $p \geq h^\vee$ and $p' \geq h^\vee$ (resp. $p' \geq h$). For $\lambda \in P_+^{p-h^\vee}$ and $\mu \in P_+^{p'-h^\vee}$ (resp. $\mu \in P_+^{\vee p'-h}$) let

$$\varphi_{\lambda,\mu}(\tau) = \frac{1}{G(\tau)} \sum_{w\in W} \varepsilon(w) q^{\frac{pp'}{2}|\frac{w(\lambda+\rho)}{p} - \frac{\mu+\rho(\text{resp.}+\rho^\vee)}{p'}|^2}. \tag{4.1.1}$$

The connection of $\varphi_{\lambda,\mu}$ to branching functions is given by

THEOREM 4.1. *Let $\mathfrak{g}$ be an affine algebra of type $X_N^{(r)}$ where either $r = r^\vee = 1$ or $r > 1$, let $\Lambda \in P_+^1$ and let $u \in \mathbb{N}$.*

(a) If $\lambda \in P_+^u$ is such that $\lambda - \Lambda \in (u-1)\Lambda_0 + Q$, then

$$b_\lambda^\Lambda(\mathfrak{g}_{[u]}) = \varphi_{\lambda,0}.$$

(b) If $k \in \mathbb{Q}$ is a principal admissible rational number with the denominator u and $\mu \in P_{u,y}^k$, $\lambda \in P_{u,y}^{k+1}$ are such that $y(\lambda^0 - \mu^0 - (u-1)\Lambda_0) - \Lambda \in Q$, then

$$b_\lambda^{\Lambda\otimes\mu} = \varphi_{\lambda^0,\mu^0}.$$

PROOF: a) (resp. b)) follows from Theorem 2.1 (resp. Theorem 3.1) and Lemma 2.2. □

Theorem 4.1b was obtained in [12, Proposition 3].

PROPOSITION 4.1. *Let $\mathfrak{g}$ be an affine algebra of type $X_N^{(r)}$ and let $\lambda \in P_+^{p-h^\vee}$, $\mu \in P_+^{p'-h^\vee}$ (resp. $\in P_+^{\vee p'-h}$). One has the following asymptotics as $\tau \downarrow 0$:*

$$\varphi_{\lambda,\mu}(\tau) \sim (pp')^{-\ell/2}|M^*/\overline{Q}^\vee|^{-1/2} \prod_{\alpha\in\overline{R}_+} 4\sin\frac{\pi(\overline{\lambda}+\overline{\rho}|\alpha)}{p} \sin\frac{\pi(\overline{\mu}+\overline{\rho}(resp.\ +\overline{\rho}^\vee|\alpha)}{p'}$$

$$\times \exp\frac{\pi i N}{12r\tau}(1-\frac{h^\vee(h^\vee(resp.\ h)+1)}{pp'}).$$

PROOF: This follows from (1.6.8), (1.6.10) and (2.2.4), using

$$\dim \mathfrak{g}(X_N) = N(h^\vee(\text{resp. } h)+1). \quad \square \tag{4.1.2}$$

Remark 4.1. Note that the group Aut $\Pi^\vee \subset$ Aut $\mathfrak{h}'$ leaves the sets $P_+^m + \mathbb{C}K$ invariant, leaves $(.|.)$ invariant, fixes ρ mod $\mathbb{C}K$ and normalizes W, and that $\varphi_{\lambda,\mu}(\tau)$ depends on λ and μ mod $\mathbb{C}K$. It follows that, defining an action of $\sigma \in$ Aut $\Pi^\vee$ on P_+ by $\sigma\sum_i m_i\Lambda_i = \sum_i m_{\sigma(i)}\Lambda_i$, we have:

$$\varphi_{\sigma(\lambda),\sigma(\mu)} = \varphi_{\lambda,\mu}.$$

4.2. Here we establish a connection between the functions $\varphi_{\lambda,\mu}$ and the characters of admissible representations. We keep distinguishing two cases as in §4.1, putting the second case in parenthesis. Let

$$\tilde{G}(\tau) = \begin{cases} G(\tau)^{h^\vee\ (\text{resp.} h)} & \text{if } A \text{ is not of type } A_{2\ell}^{(2)}, \\ (\eta(\tau/2)\eta(2\tau)\eta(\tau)^{\ell-2})^{2\ell} & \text{if} A = A_{2\ell}^{(2)}. \end{cases}$$

By [10, 4.2] we have:

$$\tilde{G}(\tau)G(\tau) = q^{|\rho|^2/2h^\vee} \prod_{\alpha\in\Delta_+\setminus\overline{\Delta}_+} (1-q^{(\Lambda_0|\alpha)})^{\text{mult}\alpha}. \tag{4.2.1}$$

PROPOSITION 4.2. *Let*

$$\Lambda = y.(\Lambda^0 - (u-1)(k+h^\vee)\Lambda_0) \in P_{u,y}^k \tag{4.2.2}$$

be a principal admissible weight and let $z \in \overline{\mathfrak{h}}$ be such that $(\alpha|z) \neq 0$ for all $\alpha \in \overline{\Delta}_+$. Then the limit

$$\psi_\Lambda(\tau) := \lim_{\varepsilon\to 0} \tilde{G}(\tau) \prod_{\alpha\in\overline{\Delta}_+} (1-e^{-2\pi i(\alpha|\varepsilon z)})\chi_\Lambda(\tau,\varepsilon z, 0) \tag{4.2.3}$$

exists and is equal to $\varphi_{\Lambda^0,\omega_{u,y}}(\tau)$ if

$$<\Lambda,\alpha> \notin \mathbb{Z} \text{ for all } \alpha \in \overline{\Delta}^\vee, \tag{4.2.4}$$

and is equal to 0 otherwise. (Recall that $\omega_{u,y}$ is defined by (1.5.16).)

PROOF: If $< \Lambda, \alpha > \in \mathbb{Z}$ for some $\alpha \in \overline{\Delta}^\vee$, then $r_\alpha \in W^\Lambda$ and hence $\sum_{w \in W^\Lambda} \varepsilon(w) e^{w(\Lambda+\rho)}(\tau, 0, 0)$

$= 0$ for all Im $\tau > 0$. Since $\prod_{\alpha \in \Delta_+ \setminus \overline{\Delta}_+} (1 - e^{-\alpha})^{\mathrm{mult}\,\alpha}(\tau, 0, 0) \neq 0$, we see from (1.6.1) that $\varphi_\Lambda(\tau) = 0$. Suppose now that (4.2.4) holds. Then we have by (4.2.1)

$$\lim_{\varepsilon \to 0} \Big(\prod_{\alpha \in \overline{\Delta}_+} (1 - e^{-\alpha}) / A_\rho \Big)(\tau, \varepsilon z, 0) = G(\tau)^{-1} \tilde{G}(\tau)^{-1}. \tag{4.2.5}$$

Let $y = t_\beta \overline{y}, \beta \in \tilde{M}, \overline{y} \in \overline{W}$. Then (1.6.1) can be rewritten as follows [13, Theorem 3.5]:

$$\chi_\Lambda(\tau, z, t) = A_{\Lambda^0 + \rho}(u\tau, \tau \overline{y}^{-1} \beta + \overline{y}^{-1} z, u^{-1}(t + (z|\beta) + \tau |\beta|^2 / 2) \Big/ A_\rho(\tau, z, t).$$

Hence, by (4.2.5),

$$\psi_\Lambda(\tau) = A_{\Lambda^0 + \rho}(u\tau, \tau \overline{y}^{-1} \beta, \tau |\beta|^2 / 2u) / G(\tau) = \varphi_{\Lambda^0, \omega_{u,y}}. \quad \square$$

Let $k \in \mathbb{Q}$ be A–admissible with the denominator $u \in \mathbb{N}$, and let

$$p = u(k + h^\vee). \tag{4.2.6}$$

The A–admissibility of k is then equivalent to

$$p, u \in \mathbb{N}, p \geq h^\vee, (p, u) = (u, r^\vee) = 1. \tag{4.2.7}$$

Denote by $\tilde{P}^k_{u,y}$ the set of all $\Lambda \in P^k_{u,y}$ satisfying (4.2.4); we shall call these principal admissible weights *nondegenerate*. Given $\overline{y} \in \overline{W}$, let $\tilde{P}^k_{\overline{y}} = \bigcup_{\beta \in \tilde{M}} \tilde{P}^k_{u, t_\beta \overline{y}}$. This set admits the following nice parameterization:

LEMMA 4.2. *Let $\lambda \in P^{p-h^\vee}_+$, $\mu \in P^{u-h^\vee}_+$ (resp. $\in P^{\vee u-h}_+$).*
(a) The map associating to the pair (λ, μ) the element

$$\Lambda_{k, \overline{y}}(\lambda, \mu) = \overline{y}(\lambda + \rho) - \frac{p}{u} \overline{y}(\mu + \rho(\text{resp.} + \rho^\vee)) + \frac{p}{u} \Lambda_0 - \rho, \tag{4.2.8}$$

establishes a bijective correspondence between the set of all such pairs and the set $\tilde{P}^k_{\overline{y}}$. In particular, $\tilde{P}^k_{\overline{y}} \neq \emptyset$ if and only if

$$k \text{ is principal admissible and } u \geq h^\vee (\text{resp. } h). \tag{4.2.9}$$

(b) Let k be principal admissible, let $\overline{y}, \overline{y}' \in \overline{W}$ and let $\overline{\sigma} = \overline{y}^{-1} \overline{y}'$. Then $\Lambda_{k, \overline{y}}(\lambda, \mu) = \Lambda_{k, \overline{y}'}(\lambda', \mu')$ if and only if

$$p^{-1}(\lambda - \overline{\sigma}.\lambda') = u^{-1}(\mu - \overline{\sigma}.\mu') := \alpha \in \tilde{M} \tag{4.2.10}$$

and

$$t_\alpha \overline{\sigma} \in \tilde{W}_+. \tag{4.2.11}$$

(c) $\tilde{P}^k_{\overline{y}}$ and $\tilde{P}^k_{\overline{y}'}$ either are disjoint or coincide, and they coincide if and only if $\overline{y}^{-1}\overline{y}' \in \overline{W}_+$.

PROOF: Let $\Lambda \in P^k_{u,y}$, $y = t_\beta \overline{y}$, $\beta \in \tilde{M}$, $\overline{y} \in \overline{W}$. Then $\Lambda = y.(\Lambda^0 - (u-1)(k+h^\vee)\Lambda_0)$ (where $\Lambda^0 \in P_+^{p-h^\vee}$) can be rewritten as follows:

$$\Lambda = \Lambda_{k,\overline{y}}(\Lambda^0, \omega_{u,y}). \tag{4.2.12}$$

If $\Lambda \in \tilde{P}^k_{\overline{y}}$, then since (4.2.4) implies (1.5.8), by Lemma 1.5, $\omega_{u,y} \in P_+^{p'-h^\vee}$ (resp. $\in P_+^{\vee p'-h}$). Conversely, this inclusion implies that Λ defined by (4.2.12) lies in $\tilde{P}^k_{\overline{y}}$. Finally Λ completely determines the pair: $\lambda = \Lambda^0$, and μ is determined since $\overline{y}$ is given. This proves (a).

Furthermore, if

$$\begin{aligned}\Lambda &= y.(\lambda - (u-1)(k+h^\vee)\Lambda_0) = \Lambda_{k,\overline{y}}(\lambda,\mu)\\ &= y'.(\lambda' - (u-1)(k+h^\vee)\Lambda_0) = \Lambda_{k,\overline{y}'}(\lambda',\mu'),\end{aligned}$$

then, by Proposition 1.5b there exists $\alpha \in \tilde{M}$, such that $y' = yt_{u\alpha}\overline{\sigma}$. Let $\sigma = t_\alpha\overline{\sigma}$. Then we have: $\sigma.\lambda' = \lambda$, hence $\sigma.\mu' = \mu$, and conditions (4.2.10 and 11) hold, and vice versa, (4.2.10 and 11) imply $\Lambda_{k,\overline{y}}(\lambda,\mu) = \Lambda_{k,\overline{y}'}(\lambda',\mu')$, proving (b). (c) follows from (b) and Proposition 1.5b. □

Finally, note

$${}^t\Lambda_{k,\overline{y}}(\lambda,\mu) = \Lambda_{k,{}^t\overline{y}}({}^t\lambda, {}^t\mu). \tag{4.2.13}$$

4.3. We turn now to the discussion of modular invariance. First, by a general result [10, Proposition 4.36] all branching functions b^Λ_λ and $b^{\Lambda\otimes\mu}_\lambda$ are holomorphic modular functions (of weight 0) in τ, $\operatorname{Im}\tau > 0$.

In order to describe the explicit transformation formulas for branching functions, we need the transformation formula for characters. Recall the action of $B = \left(\begin{smallmatrix} a & b \\ c & d\end{smallmatrix}\right) \in GL_2(\mathbb{R})$ on Y:

$$B \cdot (\tau, z, t) = (\frac{a\tau+b}{c\tau+d}, \frac{z}{c\tau+d}, (\det B)^{-1}(t - \frac{c(z|z)}{2(c\tau+d)})),$$

and its right action on functions on Y:

$$f(\tau,z,t)|_B = f(B\cdot(\tau,z,t)).$$

Recall the definitions of the congruence subgroups $\Gamma_0(n) = \{\left(\begin{smallmatrix} a & b \\ c & d\end{smallmatrix}\right) \in SL_2(\mathbb{Z}) | c \equiv 0 \mod n\}$ and the theta subgroup $\Gamma_\theta = \{\left(\begin{smallmatrix} a & b \\ c & d\end{smallmatrix}\right) \in SL_2(\mathbb{Z}) | ac$ and bd are even$\}$ of $SL_2(\mathbb{Z})$. Let $\Gamma = \Gamma_0(r)$ if $\mathfrak{g}$ is of type $X_N^{(r)} \neq A_{2\ell}^{(2)}$ and $\Gamma = \Gamma_\theta$ if it is of type $A_{2\ell}^{(2)}$. The following statement is proved in [10] for $r=1$ and is implicitly contained there in the general case.

THEOREM 4.3. *The* $\mathbb{C}$*–span of the set of normalized characters* $\{\chi_\Lambda\}_{\Lambda\in P^k_+}$ *is* Γ*–invariant.*

PROOF: In notation of [10], the space $Th_{k+h^\vee}$ is Γ–invariant by [10, Proposition 4.5a], hence the space $Th^-_{k+h^\vee}$ is Γ–invariant since the action of $GL_2(\mathbb{R})$ commutes with the action of $\overline{W}$. But $Th^-_{k+h^\vee}$ is the linear span of functions $A_{\lambda+\rho}$, $\lambda\in P^k_+$ and $\dim Th^-_{h^\vee}=1$. Together with (2.1.4) this proves the proposition. □

Let $S=\begin{pmatrix}0&-1\\1&0\end{pmatrix}$, $T=\begin{pmatrix}1&1\\0&1\end{pmatrix}\in SL_2(\mathbb{Z})$. An explicit formula for the action of S on the χ_Λ, $\Lambda\in P_+$, was found in [10], and its generalization to the case of the principal admissible weights in [13, Theorem 3.6]:

LEMMA 4.3. *Let* $\mathfrak{g}=\mathfrak{g}(A)$ *be an affine algebra of type* $X^{(1)}_\ell$ *or* $A^{(2)}_{2\ell}$. *Let* k *be a principal admissible rational number with denominator* u *and let* $\lambda\in P^k(A)$ *be a principal admissible weight. Then*

$$\chi_\lambda|_S=\sum_{\lambda'\in P^k(A)} S_{\lambda\lambda'}\chi_{\lambda'}, \tag{4.3.1}$$

where

$$\begin{aligned}S_{\lambda\lambda'}&=i^{|\overline{\Delta}_+|}u^{-\ell}(k+h^\vee)^{-\ell/2}|M^*/M|^{-1/2}\varepsilon(\overline{y}\overline{y}')\\ &\times e^{-2\pi i((\lambda^0+\rho|\beta')+(\lambda'^0+\rho|\beta)+(k+h^\vee)(\beta|\beta'))}\sum_{w\in\overline{W}}\varepsilon(w)e^{-\frac{2\pi i}{k+h^\vee}(w(\lambda^0+\rho)|\lambda'^0+\rho)}.\end{aligned} \tag{4.3.2}$$

(Here $\lambda\in P^k_{u,y}$, $\lambda'\in P^k_{u,y'}$, $y=t_\beta\overline{y}, y'=t_{\beta'}\overline{y}'$.*)* □

It is easy to check that (see (1.5.12)):

$$S_{\overset{*}{\lambda}\lambda'}=\varepsilon(\overline{w}^0)\varepsilon(\overline{w}^\lambda)\overline{S}_{\lambda\lambda'}. \tag{4.3.3}$$

Applying (4.3.1) twice and using (1.6.7) and (4.3.3), we deduce

PROPOSITION 4.3. *The matrix* $(S_{\lambda\lambda'})_{\lambda,\lambda'\in P^k(A)}$ *is a unitary (symmetric) matrix.* □

Remark 4.3. (a) Note that $\chi_\lambda|_{T^{a_0}}=e^{2\pi i m_\lambda}\chi_\lambda$, hence, by Lemma 4.3, the $\mathbb{C}$–span of the set $\{\chi_\lambda\}_{\lambda\in P^k(A)}$, which we denote by CH^k, is $SL_2(\mathbb{Z})$–invariant in the case $r=1$. Since S and T^2 generate Γ_θ, we see that CH^k is Γ_θ–invariant in the case $A^{(2)}_{2\ell}$. In the remaining cases we have [13, Theorem 3.6]:

$$\chi_\lambda|_S=\sum_{\lambda'\in P'^k(A)} S_{\lambda\lambda'}\chi'_{\lambda'}(\tau/r,z/r,t).$$

Here and further P', χ' etc. refers to the "adjacent root system" (see [10, §1.5] or [13, §3]. By Proposition 4.3, this can also be written as follows:

$$\chi'_{\lambda'}(-\frac{1}{r\tau},-\frac{z}{r\tau},t-\frac{(z/z)}{2\tau})=\sum_{\lambda\in P^k(A)}\overline{S}_{\lambda\lambda'}\chi_\lambda(\tau,z,t).$$

From these we obtain:

$$\chi_\lambda|_{ST^rS^{-1}} = \sum_{\mu\in P^k(A)}\sum_{\lambda'\in P'^k(A)} e^{2\pi i m_{\lambda'}} S_{\lambda\lambda'}\overline{S}_{\mu\lambda'}\chi_\mu .$$

Thus, CH^k is ST^rS^{-1}–invariant. Since CH^k is also T– and S^2–invariant, and since the elements ST^rS^{-1}, T and S^2 generate the group $\Gamma_0(r)$ for $r=2$ or 3, we conclude that CH^k is Γ–invariant in all cases.

(b) Let $a_0 = 1$ and denote by CH^k_y the $\mathbb{C}$–span of $\{\chi_\lambda\}_{\lambda\in P^k_{u,y}}$. Let $y = t_\beta\overline{y}$, and let $(\tau,z,t)^\wedge = u\overline{y}^{-1}t_{-\beta/u}(\tau,z/u,t/u^2)$. Then

$$STS^{-1}((\tau,z,t)^\wedge) = (ST^uS^{-1}(\tau,z,t))^\wedge .$$

Using this and (1.6.6) one derives the following transformation formula:

$$\chi_\lambda|_{ST^uS^{-1}} = \sum_{\substack{\Lambda\in P^{u(k+h^\vee)-h^\vee}\\ \mu\in P^k_{u,y}}} S_{\lambda^0\Lambda}\overline{S}_{\Lambda\mu^0}e^{2\pi i m_\Lambda}\chi_\mu .$$

Nevertheless, it is not true that CH^k_y is $\Gamma_0(u)$–invariant for each y. It is true, however, if $y=1$. To show this we use formula (1.6.6), which says that the numerator of χ_Λ, $\Lambda\in P^k_{u,y}$, is equal to

$$\hat{A}_{\Lambda+\rho}(\tau,z,t) := A_{\Lambda^0+\rho}(u\tau,\tau\overline{y}^{-1}(\beta)+\overline{y}^{-1}(z),u^{-1}(t+(z|\beta)+\tau|\beta|^2/2)).$$

Let $g=\begin{pmatrix} a & b\\ c & d\end{pmatrix}\in\Gamma_0(u)\cap\Gamma$. Then we have:

$$\hat{A}_{\Lambda+\rho}|_g = A_{\Lambda^0+\rho}(u\tau,(a\tau+b)\overline{y}^{-1}(\beta)+\overline{y}^{-1}(z),u^{-1}(t+a(\beta|z)+\frac{a(a\tau+b)|\beta|^2}{2}))\Big|_{\begin{pmatrix} a & ub\\ c/u & d\end{pmatrix}} .$$

Since Th^- is Γ–invariant, we see that $\hat{A}_{\Lambda+\rho}|_g$ is a linear combination of functions

$$A_{\mu^0+\rho}(u\tau,a\tau\overline{y}^{-1}(\beta)+\overline{y}^{-1}(z),\ u^{-1}(t+(\beta|z)+\frac{a(a\tau+b)|\beta|^2}{2})),\ \mu^0\in P^{u(k+h^\vee)-h^\vee}.$$

Assume now that $y=1$, i.e. $\beta=0$. Then $\hat{A}_{\Lambda+\rho}|_g$ is a linear combination of functions

$$A_{\mu^0+\rho}(u\tau,\overline{y}^{-1}(z),u^{-1}t) = \hat{A}_{\mu+\rho}(\tau,z,t),\ \mu\in P^k_{u,1}.$$

Note that our argument shows that CH^k_y is invariant with respect to the group $\{\begin{pmatrix} a & b\\ c & d\end{pmatrix}\in\Gamma | a\equiv 1 \mod u,\ c\equiv 0 \mod u\}$.

As in [12], Lemma 4.3 together with (3.1.2) give us a transformation law for $b^{\Lambda\otimes\mu}_\lambda$, where $\Lambda\in P^m_+$, $m\in\mathbb{N}$, and $\mu\in P^k(A)$, $\lambda\in P^{k+m}(A)$ are principle admissible weights (and $\mathfrak{g}$ is as in Lemma 4.3):

(4.3.4)
$$b^{\Lambda\otimes\mu}_\lambda(-\frac{1}{\tau}) = \sum_{\substack{\Lambda'\in P^m_+\\ \mu'\in P^k(A)\\ \lambda'\in P^{k+m}(A)}} S_{\Lambda\Lambda'}S_{\mu\mu'}\overline{S}_{\lambda\lambda'}b^{\Lambda'\otimes\mu'}_{\lambda'}(\tau).$$

Since

$$b^{\Lambda\otimes\mu}_\lambda(\tau+1) = e^{2\pi i(m_\Lambda+m_\mu-m_\lambda)}b^{\Lambda\otimes\mu}_\lambda(\tau),$$

we obtain the following corollary of (3.1.2), Remark 4.3a and Proposition 3.2:

COROLLARY 4.3. *Let $\mathfrak{g} = \mathfrak{g}(A)$ be an affine algebra of type $X_N^{(r)}$ and fix $m \in \mathbb{N}$. Then*
a) Given a principal admissible $k \in \mathbb{Q}$, the $\mathbb{C}$–span of the set

$$\{b_\lambda^{\Lambda \otimes \mu}(\tau) | \Lambda \in P_+^m,\ \mu \in P^k(A),\ \lambda \in P^{k+m}(A)\}$$

is $\Gamma_0(r/a_0)$–invariant.
b) If $u \in \mathbb{N}$ is such that $(u, h^\vee) = 1$, then the $\mathbb{C}$–span of the set

$$\{b_\lambda^\Lambda | \Lambda \in P_+^m,\ \lambda \in P_+^{um}\}$$

is $\Gamma_0(r/a_0)$–invariant.
c) Given $p, p' \in \mathbb{N}$ such that $p \geq h^\vee$, $p' \geq h^\vee$ (resp. $p' \geq h$) and $(p, p') = 1$, the $\mathbb{C}$–span of the set

$$\{\varphi_{\lambda,\mu}(\tau) | \lambda \in P_+^{p-h^\vee},\ \mu \in P_+^{p'-h^\vee}\ (\text{resp. } \in P_+^{\vee p'-h})\}$$

is $\Gamma_0(r/a_0)$–invariant. □

Proposition 2.4 shows that the conclusion of Corollary 4.3b fails if $(u, h^\vee) \neq 1$. In general, however, the subspace considered in Corollary 4.3b is at least $\Gamma_0(u)$–invariant. Indeed, let $\alpha_u = \begin{pmatrix} u & 0 \\ 0 & 1 \end{pmatrix}$. Then we have:

$$\Gamma_0(u) = SL_2(\mathbb{Z}) \cap \alpha_u^{-1} SL_2(\mathbb{Z}) \alpha_u, \tag{4.3.5}$$

$$\dot{\chi}_\lambda = \chi_\lambda|_{\alpha_u}. \tag{4.3.6}$$

Hence for $\beta = \alpha_u^{-1} \gamma \alpha_u \in \Gamma_0(u)$, where $\gamma \in SL_2(\mathbb{Z})$, we have: $\dot{\chi}_\lambda|_\beta = \chi_\lambda|_{\gamma\alpha_u}$, and by Proposition 4.3, the $\mathbb{C}$–span of the $\dot{\chi}_\lambda$ is $\Gamma_0(u)$–invariant. It follows from (2.1.8) that the $\mathbb{C}$–span of the set $\{b_\lambda^\Lambda | \Lambda \in P_+^m,\ \lambda \in P_+^{um}\}$ is $\Gamma_0(u)$–invariant.

4.4. Formula (4.3.4) and Theorem 4.1b give a transformation formula for the $\varphi_{\lambda,\mu}(\tau)$. However, a much simpler formula may be obtained by using (4.3.1) and Proposition 4.2.

PROPOSITION 4.4. *Under the assumptions of Lemma 4.3 we have:*

$$\psi_\lambda(-\frac{1}{\tau}) = (-i)^{|\overline{\Delta}_+|} \sum_{\lambda' \in P^k(A)} S_{\lambda\lambda'} \psi_{\lambda'}(\tau). \tag{4.4.1}$$

In particular, in the basis $\{\psi_\lambda\}_{\lambda \in P^k(A)}$ the transformation matrix is unitary.

PROOF: Let $\tilde{\chi}_\lambda(\tau) = \lim_{\varepsilon \to 0} \prod_{\alpha \in \overline{\Delta}_+} (1 - e^{2\pi i(\alpha|\varepsilon z)}) \chi_\lambda(\tau, \varepsilon z, 0)$. Note that (4.3.1) gives:

$$\chi_\lambda(-\frac{1}{\tau}, \frac{\varepsilon z}{\tau}, -\frac{\varepsilon^2 (z|z)}{2\tau}) = \sum_{\lambda' \in P^k(A)} S_{\lambda\lambda'} \chi_{\lambda'}(\tau, \varepsilon z, 0).$$

This can be written as follows:

$$\prod_{\alpha \in \overline{\Delta}_+} \frac{1 - e^{-2\pi i(\alpha|\varepsilon z)}}{1 - e^{-2\pi i(\alpha|\varepsilon z/\tau)}} \prod_{\alpha \in \overline{\Delta}_+} (1 - e^{-2\pi i(\alpha|\varepsilon z/\tau)}) \chi_\lambda(-\frac{1}{\tau}, \frac{\varepsilon z}{\tau}, -\frac{\varepsilon^2 (z|z)}{2\tau})$$

$$= \sum_{\lambda' \in P^k(A)} S_{\lambda\lambda'} \prod_{\alpha \in \overline{\Delta}_+} (1 - e^{-2\pi i(\alpha|\varepsilon z)}) \chi_{\lambda'}(\tau, \varepsilon z, 0).$$

Taking limit of both sides as $\varepsilon \to 0$ we obtain:

$$\tau^{|\overline{\Delta}_+|} \tilde{\chi}_\lambda(-\frac{1}{\tau}) = \sum_{\lambda' \in P^k(A)} S_{\lambda\lambda'} \tilde{\chi}_{\lambda'}(\tau). \tag{4.4.2}$$

Since, by definition (4.2.3), $\psi_\lambda(\tau) = \tilde{G}(\tau)\tilde{\chi}_\lambda(\tau)$, the proposition follows from (4.4.2) and transformation properties of $\eta(\tau)$. □

Remark 4.4. Proposition 4.4 can be extended to the remaining cases, $a_0 r = 2$ or 3, using [13, Theorem 3.7]. The result is that on the right–hand side of (4.4.1) one should add the factor $r^{|\overline{\Delta}_{+s}|}$, replace τ by τ/r, $|M^*/M|$ by $|M^*/\overline{Q}^\vee|$ and $P^k(A)$ by $P'^k(A)$. Here, as before, $P'^k(A)$ refers to the adjacent set of weights associated to the adjacent root system.

LEMMA 4.4.. *Let p and p' be relatively prime positive integers and let $\lambda \in P_+^p, \lambda' \in P_+^{p'}$ be regular weights. Then $\mu := p'\lambda - p\lambda' + p\Lambda_0$ is a regular weight.*

PROOF: We have to show that $(\mu|\alpha) \neq 0$ for any $\alpha \in \Delta_+$. Let $\alpha = \gamma + nK$, where $\gamma \in \overline{\Delta}_+$, $n \in \mathbb{Z}_+$. Then:

$$(\mu|\alpha) = p'(\lambda|\gamma) - p(\lambda'|\gamma) + pn,$$

hence p divides $(\lambda|\gamma)$ if $(\mu|\alpha) \neq 0$, which is impossible since $0 < (\lambda|\gamma) < p$ for any $\gamma \in \overline{\Delta}_+$. □

Let p, p' be relatively prime integers $\geq h^\vee$. We define a map of $P_+^{p-h^\vee} \times P_+^{p'-h^\vee}$ into itself, denoted by $(\mu, \mu') \longmapsto (\tilde{\mu}, \tilde{\mu}')$, and $\varepsilon_{\mu,\mu'}$, $\varepsilon'_{\mu,\mu'} = \pm 1$ as follows. Due to Lemma 4.4, there exist a unique $\tilde{\mu} \in P_+^{p-h^\vee}$, $\tilde{\mu}' \in P_+^{p'-h^\vee}$ and unique $w, w' \in \overline{W}$ such that

$$p'(\mu + \rho) - p(\mu' + \rho) - w(\overline{\tilde{\mu}} + \overline{\rho}) \in pM \text{ and}$$
$$p(\mu' + \rho) - p'(\mu + \rho) - w'(\overline{\tilde{\mu}}' + \overline{\rho}) \in p'M,$$

and we let

$$\varepsilon_{\mu,\mu'} := \varepsilon(w), \ \varepsilon'_{\mu\mu'} := \varepsilon(w').$$

In particular, for the pair $(\mu, p'\Lambda_0)$, we denote $\mu_{p'} = \tilde{\lambda}, \varepsilon_{p'}(\mu) = \varepsilon_{\mu,p'\Lambda_0}$.

Note also that for an integer $p \geq h^\vee$ and $\lambda, \lambda' \in P_+^{p-h^\vee}$ formula (4.3.2) turns into

$$S_{\lambda\lambda'} = i^{|\overline{\Delta}_+|} p^{-\ell/2} |M^*/M|^{-1/2} \sum_{w \in \overline{W}} \varepsilon(w) e^{-\frac{2\pi i}{p}(w(\overline{\lambda}+\overline{\rho})|\overline{\lambda}'+\overline{\rho})}. \tag{4.4.3}$$

We can rewrite now Proposition 4.4 in a form more suitable for applications.

THEOREM 4.4. *Let $\mathfrak{g} = \mathfrak{g}(A)$ be an affine algebra associated either to a symmetric matrix A or to $A = A_{2\ell}^{(2)}$. Let p and p' be relatively prime integers such that $p, p' \geq h$ and let $\lambda \in P_+^{p-h^\vee}$, $\lambda' \in P_+^{p'-h^\vee}$. Then*

$$\varphi_{\lambda,\lambda'}(-\frac{1}{\tau}) = \sum_{(\mu,\mu') \in P_+^{p-h^\vee} \times P_+^{p'-h^\vee} \mod \tilde{W}_+} S_{(\lambda\lambda')(\mu,\mu')} \varphi_{\mu\mu'}, \tag{4.4.4}$$

where

$$\begin{aligned} S_{(\lambda\lambda')(\mu\mu')} &= (-1)^{|\overline{\Delta}+|} \varepsilon_{p'}(\lambda) \varepsilon_p(\mu') |J|^{1/2} e^{2\pi i((\lambda+\rho|\mu'+\rho)+(\mu+\rho|\lambda'+\rho))} S_{\lambda_{p'}\mu} S_{\lambda'\mu'_p} \\ &= (-1)^{|\overline{\Delta}+|} |J|^{1/2} \varepsilon_{\mu\mu'} \varepsilon'_{\mu\mu'} S_{\lambda\tilde{\mu}} S_{\lambda'\tilde{\mu}'}. \end{aligned} \tag{4.4.5}$$

PROOF: Using that $\varphi_{\lambda,\lambda'} = \varphi_{\lambda',\lambda}$ we may assume that p' is odd (we shall need this in the proof for $A_{2\ell}^{(2)}$). Note that (4.2.3) can be written as follows, using (4.2.8):

$$\varphi_{\lambda,\lambda'} = \psi_{\Lambda_{k,\overline{y}}(\lambda,\lambda')}(\tau) \text{ for any } \overline{y} \in \overline{W}. \tag{4.4.6}$$

Hence we have by Proposition 4.4:

$$\varphi_{\lambda,\lambda'}(-1/\tau) = (-i)^{|\overline{\Delta}+|} \sum_{(\mu,\mu') \in P^{p-h^\vee} \times P_+^{p'-h^\vee} \mod \tilde{W}_+} \sum_{\overline{y} \in \overline{W}} S_{\Lambda_{k,1}(\lambda,\lambda'),\Lambda_{k,\overline{y}}(\mu,\mu')} \varphi_{\mu,\mu'}(\tau).$$

Hence we need to prove

$$S_{(\lambda\lambda')(\mu\mu')} = (-i)^{|\overline{\Delta}+|} S_{\Lambda_{k,1}(\lambda,\lambda'),\Lambda_{k,\overline{y}}(\mu,\mu')}. \tag{4.4.7}$$

By (4.3.2) the right–hand side of (4.4.7) is equal to

$$\begin{aligned} &\sum_{y \in \overline{W}} \varepsilon(y) (pp')^{-\ell/2} |M^*/M|^{-1/2} e^{-2\pi i((\lambda+\rho|\beta')+(\mu+\rho|\beta)+(p/p')(\beta|\beta'))} \\ &\times \sum_{w \in \overline{W}} \varepsilon(w) e^{-\frac{2\pi i p'}{p}(w(\lambda+\rho)|\mu+\rho)}, \end{aligned} \tag{4.4.8}$$

where $\beta = p'\Lambda_0 - \lambda' - \rho$, $\beta' = p'\Lambda_0 - y(\mu' + \rho)$. Substituting β and β' in (4.4.8), we obtain:

$$\begin{aligned} &(pp')^{-\ell/2} |M^*/M|^{-1/2} e^{2\pi i((\lambda+\rho|\mu'+\rho)+(\lambda'+\rho|\mu+\rho))} \\ &\times \sum_{y \in \overline{W}} \varepsilon(y) e^{-2\pi i(p/p')(y(\lambda'+\rho)|\mu'+\rho)} \sum_{w \in \overline{W}} \varepsilon(w) e^{-2\pi i(p'/p)(w(\lambda+\rho)|\mu+\rho)}. \end{aligned}$$

This gives us (4.4.4) with the first expression of $S_{(\lambda\lambda')(\mu\mu')}$ in (4.4.5). To get the second expression, we rewrite $e^{2\pi i(\lambda+\rho|\mu'+\rho)} = e^{2\pi i(w(\lambda+\rho)|\mu'+\rho)} = e^{\frac{2\pi i}{p}(pw(\lambda+\rho)|\mu'+\rho)}$, and similarly for $e^{2\pi i(\lambda'+\rho|\mu+\rho)}$. By (4.4.3), (1.3.5) and the definition of the map $(\mu, \mu') \longmapsto (\tilde{\mu}, \tilde{\mu}')$ and the $\varepsilon_{\mu\mu'}, \varepsilon'_{\mu\mu'}$, this is equal to $(-1)^{|\overline{\Delta}+|} |J|^{1/2} \varepsilon_{\mu\mu'} \varepsilon'_{\mu\mu'} S_{\lambda\tilde{\mu}} S_{\lambda'\tilde{\mu}'}$. □

Using Theorem 4.1a and the fact that $S_{00} = 1$, we obtain the following corollary of Theorem 4.4:

COROLLARY 4.4. *Let $\mathfrak{g}$ be as in Theorem 4.4 and let $u \in \mathbb{N}$ be relatively prime to h. For each $\lambda \in P_+^u$ choose the unique $\Lambda \in P_+^1$ such that $\lambda - \Lambda \in (u-1)\Lambda_0 + Q$, and let $b_\lambda(\tau) = b_\lambda^\Lambda(\tau)$. Then*

$$b_\lambda(-\frac{1}{\tau}) = \varepsilon_u(0)\varepsilon_h(\lambda)|J|^{1/2} \sum_{\mu \in P_+^u} e^{2\pi i(\lambda+\mu|\rho)} S_{\lambda_h,\mu} b_\mu(\tau). \qquad \square$$

Note that $\varepsilon_u(0)$ is the generalized Legendre symbol [7]:

$$\varepsilon_u(0) = \prod_{\alpha \in \overline{\Delta}_+} \sin \frac{\pi u(\overline{\rho}|\alpha)}{h^\vee} \Big/ \sin \frac{\pi(\overline{\rho}|\alpha)}{h^\vee}.$$

In a similar way, using Proposition 4.4 and Remark 4.4 one can derive the following transformation formula valid in all cases:

$$\varphi_{\lambda,\lambda'}(-\frac{1}{\tau}) = \sum_{\substack{\mu \in P_+^{\prime p - h^\vee} \\ \mu' \in P^{\prime p' - h^\vee} (\text{resp} \in P^{\vee p' - h})}} S_{(\lambda\lambda')(\mu\mu')} \varphi'_{\mu,\mu'}(\frac{a_0\tau}{r}), \tag{4.4.9}$$

where

$$\begin{aligned} S_{(\lambda\lambda')(\mu\mu')} &= (-1)^{|\overline{\Delta}_+|}(\frac{r}{a_0})^{|\overline{\Delta}_{+s}|}\varepsilon_{p'}(\lambda) S_{\lambda_{p'},\lambda'} \\ &\times |M^*/\overline{Q}^\vee|^{1/2}\varepsilon_p(\mu') e^{2\pi i((\lambda+\rho|\mu'+\rho')+(\mu+\rho'|\lambda'+\rho)} S_{\mu\mu'_p} \\ &(\text{resp. } \times |M^*/M|^{1/2}\varepsilon_p^\vee(\mu') e^{2\pi i((\lambda+\rho|\mu'+\rho^\vee)+(\mu+\rho|\lambda'+\rho^\vee)} S^\vee_{\mu\mu_p'^\vee}). \end{aligned}$$

Here for $\lambda \in P_+^{\vee p'-h}$ we define $\lambda_p^\vee$ and $\varepsilon_p^\vee(\lambda)$ as follows: there exists a unique $w \in \overline{W}$ and a unique $\lambda_p^\vee \in P_+^{\vee p-h}$ such that $p'(\overline{\lambda} + \overline{\rho}^\vee) - w(\overline{\lambda}_p^\vee + \overline{\rho}) \in pM$, and we let $\varepsilon_p^\vee(\lambda) = \varepsilon(w)$; for $\lambda, \lambda' \in P_+^{\vee p'-h}$ we let (cf. (4.4.3)):

$$S^\vee_{\lambda\lambda'} = i^{|\overline{\Delta}_+|} p'^{-\ell/2} |M^*/M|^{1/2} \sum_{w \in \overline{W}} \varepsilon(w) e^{-\frac{2\pi i}{p'}(w(\overline{\lambda}+\overline{\rho})|\overline{\mu}+\overline{\rho}^\vee)}.$$

In conclusion, note the following useful property of the map $(\lambda,\lambda') \longmapsto (\tilde{\lambda},\tilde{\lambda}')$:

LEMMA 4.4. *Let $p, p' \in \mathbb{N}$ be such that $p, p' \geq h^\vee$ and $(p,p') = (p, r^\vee) = (p', r^\vee) = 1$. Let $\lambda, \mu \in P_+^{p-h^\vee}$ and $\lambda', \mu' \in P_+^{p'-h^\vee}$. Then $(\tilde{\lambda},\tilde{\lambda}') = (\tilde{\mu},\tilde{\mu}')$ if and only if there exists $\sigma \in \tilde{W}_+$ such that*

$$\lambda + \rho = \sigma(\mu+\rho) \quad \textit{and} \quad \lambda' + \rho = \sigma(\mu'+\rho).$$

PROOF: The "if" part is clear by the definition. Suppose now that $\tilde{\lambda} = \tilde{\mu}$. Consider the principal admissible weights

$$\Lambda = t_\beta.(\lambda - (p'-1)(k+h^\vee)\Lambda_0), \ \Lambda' = t_{\beta'}.(\mu - (p'-1)(k+h^\vee)\Lambda_0), \tag{4.4.10}$$

where $\beta = -(\overline{\lambda}'+\overline{\rho})$, $\beta' = -(\overline{\mu}'+\overline{\rho})$. Then by the definition of $\tilde{\lambda}$ and $\tilde{\mu}$, $p'(\Lambda+\rho) \in W(\tilde{\lambda}+\rho)$ and $p'(\Lambda' + \rho) \in W(\tilde{\mu} + \tilde{\rho})$, hence

$$\Lambda' + \rho = w(\Lambda + \rho) \text{ for some } w \in W. \tag{4.4.11}$$

But the corresponding simple coroot bases are $t_{\beta'}(\Pi_{[u]})$ and $wt_{\beta}(\Pi_{[u]})$. Hence, by Proposition 1.5b, there exists $\sigma = t_\alpha \overline{\sigma} \in \tilde{W}_+$ such that

$$t_{-\beta} w^{-1} t_{\beta'} = t_{p'\alpha} \overline{\sigma}. \tag{4.4.12}$$

From (4.4.10–12) we obtain:

$$t_{p'\alpha}\overline{\sigma}.(\mu - (p' - 1)(k + h^\vee)\Lambda_0) = \lambda - (p' - 1)(k + h^\vee)\Lambda_0.$$

But the left–hand side of the last equality is equal to $\sigma.\mu - (p' - 1)(k + h^\vee)\Lambda_0$, hence $\lambda = \sigma.\mu$. Similarly, $\lambda' = \sigma'.\mu'$, and it remains to show that $\sigma = \sigma'$.

Letting $w = t_\gamma \overline{w}$, we obtain from (4.4.12):

$$t_{-\beta+\overline{w}^{-1}(\beta'-\gamma)}\overline{w}^{-1} = t_{p'\alpha}\overline{\sigma}.$$

Since $\gamma \in M$, it follows that $p'\alpha + \beta - \beta' \in M$. Since also $\beta' - \beta = \sigma'(\mu' + \rho) - (\mu' + \rho) = \overline{\sigma}'(\mu' + \rho) + p'\alpha'$, we derive that $p'(\alpha - \alpha') \in M$. Similarly, $p(\alpha - \alpha') \in M$, and since $(p, p') = 1$, we deduce that $\alpha - \alpha' \in M$. It follows that $\sigma = \sigma'$. □

References

1. Bouwknegt, P., McCarthy, J., Pilch, K., *On the free field resolutions for coset conformal field theories*, preprint, 1990.
2. Feigin, B.L., Frenkel, E.V., *Quantization of the Drinfeld–Sokolov reduction*, preprint,1990.
3. Frenkel, E.V., Kac, V.G., Wakimoto, M., In preparation.
4. Frenkel, I.B., *Representations of affine Lie algebras, Hecke modular forms and KdV type equations*, Lecture Notes in Math. 933, Springer Verlag (1982), 71–110.
5. Goddard, P., Kent, A., Olive, D., *Virasoro algebras and coset space models*, Phys. Lett. B152 (1985), 88–93.
6. Kac, V.G., *Infinite dimensional Lie algebras and Dedekind's η–function*, Funct. Anal. Appl. **8** (1974), 68–70.
7. Kac, V.G., *Simple Lie groups and the Legendre symbol*, Lecture Notes in Math 848, Springer Verlag (1981), 110–124.
8. Kac, V.G., *Infinite dimensional Lie algebras*, Progess in Math 44, Birkhäuser, Boston, 1983; Second edition:Cambridge University Press, 1985; Third edition: Cambridge University Press, 1990.
9. Kac, V.G., Peterson, D.H., *Affine Lie algebras and Hecke modular forms*, Bull. Amer. Math. Soc. **3** (1980), 1057–1061.
10. Kac, V.G., Peterson, D.H., *Infinite–dimensional Lie algebras, theta functions and modular forms*, Advances in Math. **53** (1984), 125–264.
11. Kac, V.G., Wakimoto, M., *Modular and conformal invariance constraints in representation theory of affine algebras*, Advances in Math **70** (1988), 156–236.
12. Kac, V.G., Wakimoto, M., *Modular invariant representations of infinite dimensional Lie algebras and superalgebras*, Proc. Natl. Acad. Soc. **35** (1988), 4956–4960, USA.

13. Kac, V.G., Wakimoto, M., *Classification of modular invariant representations of affine algebras,*, in Infinite–dimensional Lie algebras and groups, Adv. Ser. in Math. Phys. 7, World Sci. 1989, 138–177.
14. Lu, S., *Some results on modular invariant representations*, in: Infinite–dimensional Lie algebras and groups, Adv. Ser. in Math. Phys. 7, World Sci., 1989, 235–253.
15. Lukyanov, S.L., Fateev, V.A., Kiev, preprint, 1988.
16. Mukhi, S., Panda, S., *Fractional level current algebras and the classification of characters*, Preprint, 1990.
17. Verlinde, E., *Fusion rules and modular transformations in 2D conformal field theory*, Nucl. Phys. **B300** (1988), 360-375.

(Received: October 1990)

Acta Applicandae Mathematicae **21**: 41–56, 1990.

Computing with characters of finite groups

H. Pahlings
Lehrstuhl D für Mathematik, RWTH Aachen, Germany

Character tables of finite groups are an important tool in Representation Theory and are particularly important also for applications. The character table of a finite group (in this paper this will always mean the table of irreducible characters over **C**, in contrast to Brauer-character tables to be discussed later) contains a wealth of information on the group usually in a relatively compact form. The latter is true in particular for simple groups; even for large groups where it is completely out of the question to compute with elements or even store them in a computer, the character table might still be easily manageable. For instance the Monster group (also called Fischer Griess group or "friendly giant") has about 10^{54} elements but its character table is just a 194×194 matrix. To make full use of the information encoded in the character table it is usually necessary to perform calculations which for larger examples one certainly would not like to do without the help of a computer. So it is certainly useful to have character tables not only in book form but also stored on a computer.

In the last ten years a computer package called CAS (Character Algorithm System) has been developed in Aachen, which contains algorithms which are useful for the calculation of characters and also for extracting information out of a character table and on the other hand also a large library of character tables of finite groups. CAS has been described in the Proceedings of the Durham Conference in 1984 in a joint paper [NPP 84] with J. Neubüser and W. Plesken. But since then the system has undergone major revisions and extensions. CAS is now available on IBM's of the 370-series, in particular IBM 4341, on a VAX with VMS operating system, a Hewlett Packard HP 9000/800 and several UNIX workstations, such as Masscomp 5000 and SUN 3.

Of course, for a computer system for the handling of characters the arithmetic is crucial. Character values can be fairly large - integers with a hundred digits occur in applications - and they can be irrational. For most applications floating point numbers are useless. So CAS contains of course a multiple precision package for integers and also a module for calculating

AMS subject classifications (1980). 20C15, 20D08.
Key words. Characters, sporadic groups, combinatorics of Clifford matrices

with cyclotomic integers in an exact algebraic form. Whereas the basic data structures have not changed since [NPP 84] CAS now uses a different integral basis for the $n-th$ cyclotomic field. It turns out that for some purposes, in particular for comparing numbers in different cyclotomic fields, the standard basis $(1, \xi, \ldots, \xi^{\varphi(n)-1})$, of the $n-th$ cyclotomic field, with ξ a primitive $n-th$ root of unity and φ the Eulerian function, is not optimal, but one can find an integral basis, which better reflects the properties of the Galois groups in question, cf [Zum 89].

One of the most important new algorithms which have been implemented in CAS is the so-called LLL-algorithm due to A. K. Lenstra, H. W. Lenstra and L. Lovasz [LLL 82] for finding short vectors in lattices. It turns out that this algorithm is extremely useful when one is working with characters as it is also in many other areas. The idea of the algorithm is roughly the following:
Assume $b_1, \ldots, b_n$ is a basis for a $\mathbf{Z}$-lattice L. To initialize the algorithm the Gram-Schmidt orthogonalization procedure is used, compute

$$b_i^* = b_i - \sum_{j=1}^{i-1} \mu_{ij} b_j^* \quad 1 \leq i \leq n$$

with

$$\mu_{ij} = \frac{(b_i, b_j^*)}{(b_j^*, b_j^*)} \quad 1 \leq j \leq i-1.$$

At each level m $(2 \leq m \leq n)$ the vector b_m is reduced modulo the previous $b_j (1 \leq j \leq m-1)$ to obtain

$$\mu_{mj} \leq \frac{1}{2} \quad \text{for} \quad 1 \leq j \leq m-1.$$

In case one has $m \geq 2$ and

$$|b_m^* + \mu_{m\,m-1} b_{m-1}^*|^2 < \frac{3}{4} |b_{m-1}^*|^2$$

the vectors b_m and b_{m-1} are interchanged and m is decreased by 1.

The final output of the algorithm is an LLL-reduced basis of the lattice, i. e. one for which

$$\mu_{ij} \leq \frac{1}{2} \quad (1 \leq j < i \leq n)$$

and

$$|b_i^* + \mu_{ii-1} b_{i-1}^*|^2 \geq \frac{3}{4} |b_{i-1}^*|^2 \quad (2 \leq i \leq n)$$

holds. If $(b_1, \ldots, b_n)$ is LLL-reduced then it is known that

$$|b_j|^2 \leq 2^{i-1} |b_i^*|^2 \quad \text{for} \quad 1 \leq j \leq i \leq n,$$

$$d(L) \leq \Pi_{i=1}^n |b_i| \leq 2^{\frac{n(n-1)}{4}} d(L),$$

$$|b_1| \leq 2^{\frac{n-1}{4}} d(L)^{\frac{1}{n}},$$

where $d(L)$ is the determinant of the lattice. In many examples the norms of the vectors in an LLL-reduced basis are much smaller than those guaranteed by the above bounds.

For practical applications one has to modify the algorithm to allow it to be applied to a generating set of the lattice instead of a basis (see [Poh 87], [Sei 88]).

In the applications of the LLL-algorithm in CAS, the lattice L is usually a lattice generated by some (reducible) characters of a group G. The aim is to find irreducible characters of G or at least (generalized) characters with small norms.

Example: Assume that we want to construct the character table of J_1 just from a knowledge of the conjugacy classes and the powermaps (i. e. we assume that for each p dividing the order of J_1 and for each conjugacy class C_i of J_1 we know in which class the $p-th$ powers of the elements of C_i are). Then one can compute the characters induced from all the linear characters of the cyclic subgroups of J_1; in CAS the command "induce cyclic" with the parameter "all" does this, (without "all" one would just induce the trivial characters, which is useful for instance if one wants to compute the Artin-exponent of a group for which the character table is known). Of course, in our example it is sufficient to induce up the linear characters of the maximal cyclic subgroups. One thus obtains 32 reduced characters with norms ranging from 487 up to 43920 and, of course, the regular character. Applying the LLL-algorithm one obtains after 11 characters have been processed the first two irreducible characters and shortly afterwards another 5. Adding the trivial character to the list of irreducible characters and reducing out another application of the LLL-algorithm gives the complete character table of J_1. The total cpu-time on a microcomputer (HP 9000/800) is less then 30 seconds.

Of course this is not a very typical example for applying the LLL-algorithm. In fact it is not very common that every irreducible character of a group is an integral linear combination of characters induced from cyclic subgroups and the trivial character. Perhaps a better example would have been the construction of the character tables of the orthogonal group $0^-_{12}(2).2$ with 375 conjugacy classes in a joint work with J. S. Frame. Actually J. S. Frame computed the centralizer orders and powermaps of this group and also a dozen or so characters by hand and the rest was computed using CAS making heavy use of the LLL-algorithm. The largest successful application we had was a lattice spanned by 180 reducible characters which turned out to be 150 dimensional, where the algorithm produced almost 100 irreducible characters in one run. Of course often the LLL-algorithm cannot produce any irreducible character - this also happened in the example of $0^-_{12}(2).2$ - simply because the $\mathbf{Z}$ -span of the characters considered does not contain any norm 1 character. Then, although one may have obtained a number of characters of small norms one is faced with the task to decide whether one of the characters or perhaps some combination thereof is in fact a multiple of an irreducible character, which is a much harder task for generalized characters than it is for ordinary characters, where there are backtrack algorithms ("square dancing") cf. [Con 84], [NPP 84]. So, in such a situation it is good if one can return to ordinary characters. If one knows that the group in question does not have any symplectic characters (i. e. any irreducible character χ with Frobenius-Schur indicator $\nu_2(\chi) = \frac{1}{|G|}\sum_{g\in G}\chi(g^2)$ equal to -1), as one does for example for the orthogonal groups as discussed above, then one can conclude e. g. that any generalized character of norm 2 with Frobenius-Schur-indicator also equal to 2 must in fact be an ordinary character.

There is one case, noted by R. Parker, in which working with generalized characters one can decide, that a character in the lattice is twice a character of norm 1 (i. e. an irreducible character or the negative of it) and thus obtaining irreducible characters which are not neccessarily in the lattice itself. This is the case when the reducible characters in question form a D_n-lattice for $n \geq 5$, that is if their matrix $((\chi_i, \chi_j))$ of scalar products looks like this

$$\begin{pmatrix} 2 & -1 & & & & \\ -1 & 2 & -1 & & & \\ & -1 & 2 & & & \\ & & & 2 & -1 & -1 \\ & & & -1 & 2 & \\ & & & -1 & & 2 \end{pmatrix}.$$

Then one knows from the properties of the D_n-lattice, that $\chi_n + \chi_{n-1}$ is twice a vector of norm 1.

For $n = 4$, which unfortunately seems to be the case that arises most frequently in examples, one cannot assert this (because of the symmetry of the D_4-graph) but nevertheless one can assert that

$$\chi_1 + \chi_3 \quad \text{or} \quad \chi_1 + \chi_4 \quad \text{or} \quad \chi_3 + \chi_4$$

is twice a norm 1 character. In fortunate cases one might be able to exclude two of the possibilities by examining the characters.

Example: Trying to compute the irreducible characters of the Mathieu group M_{22} from a knowledge of the powermaps only one obtains after inducing up the linear characters of the cyclic subgroups and reducing with the trivial character that the LLL-algorithm produces just one further irreducible character (of degree 231) and a list of 10 generalized characters of norm 2, 3 and 4. If one computes the skew-symmetrized squares of these generalized characters and applies the LLL-algorithm again, one obtains 3 further irreducible characters (of degree 280, 280 and 385) and reducible generalized characters $Y_1, \ldots, Y_7$ of norm 2 and 3. From the matrix of scalar products of the first four of these

$$\begin{pmatrix} 2 & -1 & 1 & 1 \\ -1 & 2 & 0 & -1 \\ 1 & 0 & 2 & 1 \\ 1 & -1 & 1 & 2 \end{pmatrix}$$

one finds that the first 4 generate a D_4 lattice. In fact the correspondence is

$$\begin{array}{ccccc} -Y_1 & & & & \\ & \searrow & & & \\ & & Y_4 & — & (Y_3 - Y_4) \; . \\ & \nearrow & & & \\ -(Y_2 + Y_4) & & & & \end{array}$$

Now, since $Y_1(1) = 0$, $(Y_2 + Y_4)(1) = 189$, $(Y_3 - Y_4)(1) = 231$ it follows that only the sum of the last two generalized characters can be twice an irreducible. Reducing with this newly found irreducible character one obtains three further irreducible characters and one is left with three generalized characters with matrix of scalar products

$$\begin{pmatrix} 2 & 1 & -1 \\ 1 & 2 & -1 \\ -1 & -1 & 2 \end{pmatrix}.$$

Since one knows that there are exactly three irreducible characters missing (in fact by Artin's Theorem at any point in the process described one is sure that the table of reducible generalized characters contains exactly as many characters as there are irreducible characters still missing, because the LLL-algorithm always produces linearly independent generalized characters) one can easily see that the sum of these 3 generalized characters must be twice a norm 1 character. This, of course, then gives also the last irreducible characters missing.

If one wants to compute the character table of a group G with a (possibly large) normal subgroup N, where the character table of $H = G/N$ is known, one is well advised to use the method of Clifford matrices as introduced by B. Fischer (cf. [Fis 82], [Fis 85]). To utilize these methods several programs have been included in CAS, see [Bon 88]. We briefly outline this method and demonstrate its use by an example.

Keeping the notation from the previous paragraph, let $\lambda_1, \ldots, \lambda_n$ be representatives of the G-orbits of irreducible characters of N, with $\lambda_1 = 1$, the trivial character. Let $\tilde{T}_m$ be the inertia subgroup of λ_m $(1 \le m \le n)$ in G

$$\tilde{T}_m = \{g \in G \mid \lambda_m(gxg^{-1}) = \lambda_m(x) \text{ for all } x \in N\}.$$

Then the set $Irr(G)$ of irreducible characters of G partitions as follows

$$Irr(G) = \bigcup_{m=1}^{n} Irr(G, \lambda_m)$$

with

$$Irr(G, \lambda_m) := \{\chi \in Irr(G) \mid (\chi_N, \lambda_m) \neq 0\} = \{\varphi^G \mid \varphi \in Irr(\tilde{T}_m, \lambda_m)\},$$

according to Clifford's Theorem (see e.g. [Isa 76] (6.11)); here of course $Irr(\tilde{T}_m, \lambda_m) = \{\varphi \in Irr(\tilde{T}_m) \mid (\varphi_N, \lambda_m) \neq 0\}$. Any $\varphi \in Irr(\tilde{T}_m, \lambda_m)$ is of the form $\varphi = \tilde{\lambda}_m \psi$, where $\tilde{\lambda}_m$ is an extension of λ_m to a (in general) projective character of $\tilde{T}_m$ with factor set α_m (which is constant on cosets of N and can thus be considered as a factor set of $\tilde{T}_m/N$) and ψ is a projective character of $T_m = \tilde{T}_m/N$ with factor set α_m^{-1} (see e.g. [Isa 76] (11.2)).

Thus for $\chi \in Irr(G, \lambda_m)$ and $x \in G$

$$\begin{aligned} \chi(x) &= \varphi^G(x) = \textstyle\sum_k \frac{|C_G(x)|}{|C_{\tilde{T}_m}(h_k)|}\varphi(h_k), \\ &= \textstyle\sum_k \frac{|C_G(x)|}{|C_{\tilde{T}_m}(h_k)|}\tilde{\lambda}_m(h_k)\psi(h_k N) \end{aligned}$$

where the h_k are representatives of the $\tilde{T}_m$-conjugacy classes into which $x^G \cap \tilde{T}_m$ splits. If the conjugacy class x^G of x does not meet $\tilde{T}_m$, then $\chi(x)$ is, of course, zero. Now the philosophy is, not to consider the inertia group $\tilde{T}_m$ but the inertia factor $T_m = \tilde{T}_m/N$ which is after all a subgroup of H which was supposed to be known and manageable. So one groups together in the above sum those summands for which the elements $h_k N$ are conjugate in the inertia factor T_m and obtains

$$\chi(x) = \sum_e c_{ex}\psi(y_e N) \qquad (*)$$

where the $y_e N$ are now representatives of the T_m-conjugacy classes into which $(xN)^H \cap T_m$ splits (and which are α_m^{-1}-regular, for otherwise $\psi(y_e N)$ is zero, anyhow), and

$$c_{ex} = \sum_k \frac{|C_G(x)|}{|C_{\tilde{T}_m}(h_k)|}\tilde{\lambda}_m(h_k)$$

with the sum extending over all those k for which h_kN is conjugate in T_m to a particular y_eN. Observe, that the y_eN do not depend on the G-conjugacy class in which x lies but only on the G/N-conjugacy class of xN for which x_iN may be a fixed representative (assume that $x_1N, \ldots, x_rN$ are chosen representatives of the conjugacy classes of H) and of course on the intertia factor T_m. So a proper notation for y_eN is y_{ie}^mN. On the other hand c_{ex} defined above does depend on the G-conjugacy class of x; if x_{ij} is a fixed representative of the G-conjugacy class of x (the index i should remind you that $x_{ij}N$ is conjugate to x_iN in $G/N = H$), then we write

$$c^i_{(m,e)j} := c_{ex}.$$

The matrix containing this (algebraic) value in row $\alpha = (m, e)$ (ordered, say lexicographically) and column j is called the Clifford matrix $Cl(i)$ (or Fischer matrix) corresponding to the class of $x_iN \in H$. Formula (*) then can be written as

$$\chi(x_{ij}) = \sum_e c^i_{(m,e)j}\psi(y^m_{ie}N) \qquad (**)$$

if $\chi \in Irr(G, \varphi_m)$, where the sum extends over a set of representatives $y^m_{ie}N$ for the α_m^{-1}-regular conjugacy classes of T_m which fuse into the H-conjugacy class of x_iN.

The Clifford matrices satisfy orthogonality relations which follow from the usual orthogonality relations for character tables. Keeping the notation as above, let

$$\begin{array}{lcll} m^i_j &:=& [N_G(x_iN) : C_G(x_{ij})] = |N|\frac{|C_{G/N}(x_iN)|}{C_G(x_{ij})|} & , \\ c^i_j &=& |C_G(x_{ij})| & , \\ b^i_\alpha &:=& b^i_{(m,e)} := |C_{T_m}(y^m_{ie}N)| & ; \end{array}$$

so, in particular one has

$$\begin{array}{lcll} b^i_1 &:=& b^i_{(1,1)} = |C_H(x_iN)| & \text{and} \\ & & m^i_j \cdot c^i_j = |N|b^i_1 & \text{, independent of } j. \end{array}$$

Then one has

Orthogonality relations:

$$\sum_\alpha b^i_\alpha c^i_{\alpha j}\overline{c^i_{\alpha k}} = \delta_{jk}c^i_j.$$

As a consequence one finds that the Clifford matrix $Cl(i)$ is always a square matrix. Also one obtains orthogonality relations for the rows of the Clifford matrices:

$$\sum_j m_j^i c_{\alpha j}^i \overline{c_{\beta j}^i} = \delta_{\alpha\beta} \frac{b_1^i}{b_\alpha^i} |N|.$$

The method of Clifford matrices works particularly smoothly for abelian N and if one has a split extension (i.e. the normal subgroup N has a complement in G). Then one has to deal only with ordinary characters instead of projective ones. Also the rows of the Clifford matrices are then $N_G(x_i N)$-orbit sums of characters of N/N_{x_i}, where $N_{x_i} = [x_i, N]$. In particular the first column is known:

$$c_{(m,e)j}^i = b_{(1,1)}^i \left(b_{(m,e)}^i\right)^{-1}.$$

Example: Let us consider a split extension $G = 2^4 : A_5$ of $N = \mathbf{Z}_2^4$ by $H = A_5$ with non-trivial action of H on N. (In fact it can be shown that any such extension will be split (see e.g. [HoP 89]), so our assumption is not restrictive.) Since A_5 has no non-trivial representation of degree less than 4 over $GF(2)$ the action of H on N is irreducible and it follows immediately that

H has one orbit (of length 15) on $N \backslash \{1\}$ or
H has two orbits (of lengths 5 and 10) on $N \backslash \{1\}$.

Let us consider the second case (the first one being almost trivial). It follows that H has three orbits on $Irr(N)$ with stabilizers $T_1 = H = A_5$, $T_2 \neq A_4$, $T_3 \neq S_3$. The first point is, to write down the character tables of these groups and determine the fusion maps of T_i in H:

	x_1	x_2	x_3	x_4	x_5	representative
	$1A$	$2A$	$3A$	$5A$	$5B$	CAS notation
$T_1 = H$	60	4	3	5	5	centralizer order
$\psi_1^{(1)}$	1	1	1	1	1	
$\psi_2^{(1)}$	3	-1	0	A	A^*	$A = \frac{1}{2}(1+\sqrt{5})$
$\psi_3^{(1)}$	3	-1	0	A^*	A	$A^* = \frac{1}{2}(1-\sqrt{5})$
$\psi_4^{(1)}$	4	0	1	-1	-1	
$\psi_5^{(1)}$	5	1	-1	0	0	

$T_2 = A_4$	y_{11}^2	y_{21}^2	y_{31}^2	y_{32}^2	representative
	$1A$	$2A$	$3A$	$3B$	CAS notation
	12	4	3	3	centralizer order
$\psi_1^{(2)}$	1	1	1	1	
$\psi_2^{(2)}$	1	1	B	$\overline{B}$	
$\psi_3^{(2)}$	1	1	$\overline{B}$	B	
$\psi_4^{(2)}$	3	-1	0	0	

$$B = -\tfrac{1}{2}(1 - \sqrt{-3}).$$

$T_3 = S_3$	y_{11}^3	y_{21}^3	y_{31}^3
	$1A$	$2A$	$3A$
	6	2	3
$\psi_1^{(3)}$	1	1	1
$\psi_2^{(3)}$	1	-1	1
$\psi_3^{(3)}$	2	0	-1

In these tables $x_iN = y_{i1}^1 N$ and $y_{ie}^m N$ have been replaced by x_i and y_{ie}^m respectively.
The fusion maps are given by the names of the representatives:
remember $y_{ie}^m N$ of T_m always fuses to x_iN of H.
Thus one has 5 Clifford matrices corresponding to the classes of H. These are generated by the CAS-command **clmake** together with the weights $b_{m,e}^i$ for the rows, the command **clsplit** computes then the first columns. The result in our example is, of course

$Cl(1)$:

$b_{(m,e)}^1$	x_{11}	x_{12}	x_{13}
60	1	1	1
12	5		
6	10		

$Cl(2)$:

$b_{(m,e)}^2$	x_{21}	x_{22}	x_{23}
4	1	1	1
4	1		
2	2		

$Cl(3)$:

$b_{(m,l)}^3$	x_{31}	x_{32}	x_{33}	x_{34}
3	1	1	1	1
3	1			
3	1			
3	1			

$Cl(4)$:

$b_{(m,l)}^4$	x_{41}
5	1

$Cl(5):$

$b^5_{(m,l)}$	x_{51}
5	1

Denoting the unknown elements in the second and third row of $Cl(1)$ by x, y and z, u respectively, the orthogonality relations for the columns give immediately $x + z = -1$ and $y + u = -1$ and $c^1_2 = 60 + 12x^2 + 6(x+1)^2$. Observe that since x is the value of an orbit sum of (5) irreducible characters of an elementary abelian 2 group, x must be a rational integer, in fact x must be one of the numbers $3, 1, -1, -3$ by the same reason. Since c^1_2 must be a divisor of $|G| = 16 \cdot 60$, only $x = 1$ and $x = -3$ are possible. The same holds true for y. By orthogonality the values of x and y must be different and we may arrange our columns so that $x = -3$ and $y = 1$.
Similarly an appeal to the fact that the rows of the Clifford matrices are orbit sums of characters and the orthogonality relations immediately give the missing values in $Cl(2)$ and $Cl(3)$. As one can see, Clifford matrices in the split case are often fully determined by their first columns and hence by the inertia factor groups. So one can store a library of such Clifford matrices which then can be used in many examples, thus automatizing the computations. This is not yet implemented in the current version of CAS.

Now the centralizer orders c^i_j can be computed by the orthogonality relations and also the weights m^i_j (command: **clcalc**). We reproduce the complete Clifford matrices:

$b^1_{(m,e)}$	x_{11}	x_{12}	x_{13}
60	1	1	1
12	5	-3	1
6	10	2	-2
m^1_j	1	5	10
c^1_j	960	192	96

$b^2_{(m,e)}$	x_{21}	x_{22}	x_{23}
4	1	1	1
4	1	1	-1
2	2	-2	0
m^2_j	4	4	8
c^2_j	16	16	8

$b^3_{(m,e)}$	x_{31}	x_{32}	x_{33}	x_{34}
3	1	1	1	1
3	1	1	−1	−1
3	1	−1	1	−1
3	1	−1	−1	1
m^3_j	4	4	4	4
c^3_j	12	12	12	12

$b^4_{(m,e)}$	x_{41}
5	1
m^4_j	16
c^4_j	5

$b^5_{(m,e)}$	x_{51}
5	1
m^5_j	16
c^5_j	5

Now the entries of the character table of $G = 2^4 : A_5$ can be computed using (**). This can be viewed as a matrix product, the first factor M_1 consisting of the diagonal sum of the character tables of the various inertia factors $T_1 = H, T_2, T_3, \ldots$ and the second M_2 containing the Clifford matrices suitably arranged. In our example

$$M_2 = \left[\begin{array}{cccccccccccc}
1 & 1 & 1 & & & & & & & & & \\
 & & & 1 & 1 & 1 & & & & & & \\
 & & & & & & 1 & 1 & 1 & 1 & & \\
 & & & & & & & & & & 1 & \\
 & & & & & & & & & & & 1 \\
\hline
5 & -3 & 1 & & & & & & & & & \\
 & & & 1 & 1 & -1 & & & & & & \\
 & & & & & & 1 & 1 & -1 & -1 & & \\
 & & & & & & 1 & -1 & 1 & -1 & & \\
\hline
10 & 2 & -2 & & & & & & & & & \\
 & & & 2 & -2 & 0 & & & & & & \\
 & & & & & & 1 & -1 & -1 & 1 & &
\end{array}\right]$$

Then the matrix of character values for G is $M_1 \cdot M_2$.

	x_{11}	x_{12}	x_{13}	x_{21}	x_{22}	x_{23}	x_{31}	x_{32}	x_{33}	x_{34}	x_{41}	x_{51}
χ_1	1	1	1	1	1	1	1	1	1	1	1	1
χ_2	3	3	3	-1	-1	-1	0	0	0	0	A	A^*
χ_3	3	3	3	-1	-1	-1	0	0	0	0	A^*	A
χ_4	4	4	4	0	0	0	1	1	1	1	-1	-1
χ_5	5	5	5	1	1	1	-1	-1	-1	-1	0	0
χ_6	5	-3	1	1	1	-1	2	0	0	-2	0	0
χ_7	5	-3	1	1	1	-1	-1	C	$-C$	1	0	0
χ_8	5	-3	1	1	1	-1	-1	$-C$	C	1	0	0
χ_9	15	3	3	-1	-1	1	0	0	0	0	0	0
χ_{10}	10	2	-2	2	-2	0	1	-1	-1	1	0	0
χ_{11}	10	2	-2	-2	2	0	1	-1	-1	1	0	0
χ_{12}	20	4	-4	0	0	0	-1	1	1	-1	0	0

where $C = \sqrt{-3}$ and $A = \frac{1}{2}(1+\sqrt{5})$ as before.
CAS does this computation of course automatically using **clproduct** for the various inertia factors. What still remains to do is to compute the element orders of the classes and the fusion maps. Some of the element orders are clear from the beginning: $o(x_{i1}) =$ order of x_iN in H, since we have a split extension, and $o(x_{1j}) = 2$ for $j > 2$, since $x_{1j} \in N$ and N is elementary abelian in our case. This information is enough to find the skew-symmetrized square $\chi_6^{[1^2]}$ of, say χ_6, which is defined by

$$\chi^{[1^2]}(g) = \frac{1}{2}(\chi(g)^2 - \chi(g^2)).$$

In fact, the first four character values of χ_6 are sufficient to conclude that $\chi_6^{[1^2]} = \chi_{11}$. This in turn implies that the squares of $x_{22}, x_{23}, x_{32}, x_{33}, x_{34}$ are conjugate to $x_{12}, x_{13}, x_{31}, x_{31}, x_{31}$, respectively, hence
$o(x_{12}) = o(x_{23}) = 4$, $o(x_{32}) = o(x_{33}) = o(x_{34}) = 6$. Thus the element orders and also the second powermap are found. The other powermaps are trivially obtained.

In the above example it was very easy to see that a character $(\chi_6^{[1^2]})$ was induced from a character of one of the inertia subgroups $(\tilde{T}_3)$. In general it is very useful that one can decompose a (reducible) character into a sum of characters which are induced up from inertia subgroups, by forming scalar products of a character χ of G with the rows $c^i_{(m,l)}$ of the Clifford matrices

$Cl(i)$. These are defined by

$$(\chi, c^i_{(m,e)}) := \frac{b^i_{(m,e)}}{b_{(1,1)}|N|} \sum_j m^i_j \chi \frac{x_{ij}}{c_{(m,e)j}}$$

Now, if

$$\chi = \sum_{m=1}^{n} (\psi^{(m)})^G$$

with

$$\psi^{(m)} = \sum_k a^{(m)}_k \varphi^{(m)}_k, \quad \varphi^{(m)}_k \in Irr(\tilde{T}_m, \lambda_m)$$

then

$$(\chi, c^i_{(m,e)}) = \sum_k a^{(m)}_k \psi^{(m)}_k (y^m_{ie} N) \quad (\psi^{(m)}_k \in Irr(T_m))$$

Thus using these scalar products, which can be computed in CAS using the **cldecompose** command, one can decompose a character into its pieces belonging to the various inertia groups.

For a generalization of the method to generic character tables cf [PaP 88].

The methods described above have been useful for computing the character tables of most of the maximal subgroups of the sporadic simple groups (excluding those of the Monster and the Baby-Monster for the moment). These character tables in turn have proved useful for many applications. One of the applications concerns the inverse problem of Galois-Theory. Using a theorem of Belyi, Fried, Matzat and Thompson (see e. g. [Mat 87], [Mat 88], [Ser 88]) it was proved by various authors, that all the sporadic simple groups are Galois groups over **Q** with the possible exception the Mathieu group M_{23}. In all cases the character table of the simple group were used and in several cases it was also essential to consider also the character tables of maximal subgroups, see also [Pah 90].

The character tables of maximal subgroups of the sporadic simple groups have also proved to be valuable for the computation of the decomposition numbers (or, equivalently, the Brauer - character tables) of the sporadic simple groups. In fact, a standard method for computing decomposition numbers is to produce as many projective characters (i. e. characters of projective modules) as possible and one way to obtain these is to induce up projective characters of maximal subgroups.

References

[Bon 88] O. Bonten, Clifford-Matrizen, Diplomarbeit Aachen 1988

[Con 84] J. H. Conway, Character Calisthenics; in: Computational Group Theory (ed. M. D. Atkinson) pp. 249-266, London: Academic Press 1984

[Fis 82] B. Fischer, Clifford Matrizen, manuscript 1982

[Fis 85] B. Fischer, unpublished manuscript 1985

[HoP 89] D. F. Holt, W. Plesken, Perfect Groups.
Oxford: Clarendon Press 1989

[Isa 76] I. M. Isaacs, Character Theory of Finite Groups.
New York: Academic Press 1976

[LLL 82] A. K. Lenstra, H. W. Lenstra, L. Lovasz, Factoring polynomials with rational coefficients. Math. Ann. **261** (1982), 513-534.

[Mat 87] B. H. Matzat, Konstruktive Galoistheorie. (Lecture Notes in Mathematics 1284) Springer 1987, x + 286 pp

[Mat 88] B. H. Matzat, Über das Umkehrproblem der Galoischen Theorie. J. ber. d. Dt. Math. Verein. **90** (1988), 155-183

[NPP 84] J. Neubüser, H. Pahlings, W. Plesken, CAS; design and use of a system for the handling of characters of finite groups,
in: Computational Group Theory (ed. M. D. Atkinson) pp 195-247,
London: Academic Press 1984

[Pah 90] H. Pahlings, Realizing finite groups as Galois groups. To appear

[PaP 87] H. Pahlings, W. Plesken, Group actions on Cartesian powers with applications to representation theory.

J. reine angew. Math., **380** (1987), 178-195

[Poh 87] M. Pohst, A modification of the LLL reduction algorithm, J. Symbolic Computation **4** (1987), 123-127

[Sei 88] C. Seidler, der LLL-Algorithmus und seine Implementation in CAS. Diplomarbeit, Aachen 1988

[Ser 88] J.-P. Serre, Groupes de Galois sur Q. Séminaire Bourbaki no. 689, 1987-88, Astérisque 161-162, (1988), 73-85

[Zum 89] M. Zumbroich, Grundlagen einer Arithmetik in Kreisteilungskörpern und deren Implementation in CAS. Diplomarbeit, Aachen 1989

(Received: October 1990)

Acta Applicandae Mathematicae **21**: 57–76, 1990.

Some remarks on the computation of complements and normalizers in soluble groups

F. Celler, J. Neubüser, C.R.B. Wright
Lehrstuhl D für Mathematik, RWTH Aachen, Germany

Dedicated to Wolfgang Gaschütz, to whom we owe so much of our understanding of soluble groups, on the occasion of his seventieth birthday

1. Introduction.

The three lectures "Computing in Soluble Groups" given by the second author as part of the "Seminar on Computational Algebra" at the Department of Mathematics of the University of Rome "Tor Vergata" on March 22./23., 1990 embedded a report on some recent work on the topic mentioned in the title of this paper into a survey of the state of the art in that wider field. In this paper a reminiscence of that overview will be confined to some very sketchy historical remarks in this introduction, but the bibliography should still be sufficient to obtain a reasonable coverage of the literature on the field by tracing back.

For investigations that involve calculation with the elements of a finite group the elements must be represented in a form that allows efficient multiplication, inversion and, last but not least, comparison. The first programs of that kind, for the determination of the subgroup lattice, already used both permutations for arbitrary groups and, for p-groups, ordered words in a generating set corresponding to a central series with cyclic factors [Neu 61]. The systematic use in such programs of basic notions from permutation group theory, in particular the stabilizer chain, was initiated by C. Sims in three fundamental papers in the late sixties and made generally available with many refinements and extensions in the precursors and early versions of the CAYLEY system (cf. [Can 84] for a comprehensive overview). On the other hand a systematic use of the properties of composition and chief series of finite soluble groups for the investigation of their structure only started with an algorithm for the determination of conjugacy classes of p-groups [FeN 79] and the subsequent development of the SOGOS system [LNS 84], preceded by the invention [Mac 74], further development, and very successful application [New 76] of the Nilpotent Quotient Algorithm, which constructs central series of p-factor groups of a finitely presented group. Computing methods for soluble groups have recently received

AMS subject classification (1980). 20D10.
Key words. Soluble groups, normal complements, computation of normalizers.

much interest: collection methods for the multiplication of the ordered words mentioned above that had been believed to be fairly well understood for some time [HaN 76] have been reinvestigated with remarkable success [L-GS 90][Va-L 90], a number of rather powerful new algorithms have been described [GlS 90], [MeN 89], [Con 90], applications to explicit classifications of p-groups have been made [O'Br 90], p-group and soluble group techniques have entered into more general situations [Hol 84], [Sim 90b], and a number of proposals have been made for the generalisation of the NQA to a "Soluble Quotient Algorithm" [Ple 87], [Sim 90a].

In this paper we describe methods for calculating the classes of complements of a given normal subgroup of a soluble group and their connection to the calculation of the normalizer of a given subgroup. These methods can in some sense, on which we comment at the end of the paper, be viewed as counterparts to the algorithms for the determination of conjugacy classes of elements given in [MeN 89]. They supplement the algorithm for the determination of the normalizer described in [GlS 90] in those cases in which the latter has to fall back on the general "orbit-stabilizer" algorithm outlined in section 3 of [LNS 84]. The algorithms described in this paper have been implemented by the first author as part of the GAP system [NNS 88], which, among many other things, contains in a much better organized form almost all that had previously been incorporated into SOGOS.

2. Basic definitions.

To make the paper reasonably self-contained we repeat once more the basic definitions, referring to [LNS 84] for details.

Let $G = G_0 > \ldots > G_n = \langle 1 \rangle$ be a composition series of the finite soluble group G with factors $G_{i-1}/G_i = \langle g_i G_i \rangle$ of order some prime p_i. Then $\langle g_1, \ldots, g_n \rangle$ is a generating sequence, called a *PAG sequence*, with a defining set of relations

$$\begin{array}{rcll} g_i^{p_i} & = & w_{ii}(g_{i+1}, \ldots, g_n) & \text{for } 1 \leq i \leq n \\ [g_i, g_j] & = & w_{ij}(g_{j+1}, \ldots, g_n) & \text{for } 1 \leq j < i \leq n. \end{array}$$

We may assume that in addition a subsequence $N_1 = G_0, N_2 = G_{i_2}, \ldots, N_j = G_{i_j}, \ldots, N_m = G_n$ of the composition series forms a normal series with elementary abelian factors, or even a chief series of G.

Each element of G can be uniquely expressed in the form

$$g = g_1^{\nu_1} \dots g_n^{\nu_n} \quad \text{with } 0 \le \nu_i < p_i,$$

and multiplication can be performed by a collection process using the relations.

If $\nu_i = 0$ for $i = 1, \dots, k-1$ and $\nu_k \neq 0$, we call $\nu_k =: \lambda(g)$ the *leading exponent* and $k =: w(g)$ the *weight* of g. With respect to $(g_1, \dots, g_n)$ each nontrivial subgroup $U \le G$ has a unique *canonical generating sequence* $(u_1, \dots, u_s)$, abbreviated CGS, with the following properties

(1)	$(u_1, \dots, u_s)$	is a PAG sequence for U,
(2)	$w(u_i) > w(u_j)$	for $i > j$,
(3)	$\lambda(u_i) = 1$	for $i = 1, \dots, s$,
(4)	$\nu_{w(u_i)}(u_j) = 0$	for $i \neq j$.

Given any generating set of a subgroup $U \le G$, its CGS can be determined by a "noncommutative Gauß algorithm".

The use of "homomorphism principles" has been recommended in section 3 of [LNS 84]. We shall apply them in the following form.

Let G be a group acting on a set Ω and let $N \lhd G$. Let $\omega \in \Omega$. Then an orbit ω^N is a block for G. Let $Stab_G(\omega)$ be the stabilizer of ω and $Stab_G(\omega^N) = \{g \mid g \in G, \omega^g \in \omega^N\}$ the block stabilizer of ω^N. For each $b \in Stab_G(\omega^N)$ there exists $n_b \in N$ such that $\omega^b = \omega^{n_b}$, that is, $bn_b^{-1} \in Stab_G(\omega)$. Hence with $Stab_{G/N}(\omega^N) = \langle b_1 N, \dots, b_k N\rangle$ and $Stab_N(\omega) = \langle n_1, \dots, n_l\rangle$ we have

$$Stab_G(\omega) = \langle b_1 n_{b_1}^{-1}, \dots, b_k n_{b_k}^{-1}, n_1, \dots, n_l\rangle.$$

In our applications we will be able to compute $Stab_{G/N}(\omega^N)$ and for given $bN \in Stab_{G/N}(\omega^N)$ compute an element n_b such that $bn_b^{-1} \in Stab_G(\omega)$.

3. The first cohomology group.

Let G be an arbitrary group with finite presentation

$$G = \langle g_1, \dots, g_n \mid R_j(g_1, \dots, g_n) = 1, \ j = 1, \dots, r\rangle,$$

let M be an elementary abelian p-group of rank d on which $g \in G$ acts by $g : m \to m^g$. As is well known, the group of 1-cocycles can be defined as

$$Z^1 = \{\gamma : G \to M \mid \ \gamma(gg') = \gamma(g)^{g'} \gamma(g'), \ \ \forall g, g' \in G\},$$

the group of 1-coboundaries as

$$B^1 = \{\gamma_m : G \to M \mid \exists m \in M, \ \forall g \in G, \ \gamma_m(g) = m m^{-g}\},$$

and the 1-cohomology group $H^1(G, M) := Z^1/B^1$.

The definition of Z^1 makes clear that γ is uniquely determined by its values on the generators $g_1, \ldots, g_n$ of G, i.e. that $\beta : \gamma \to (\gamma(g_1), \ldots, \gamma(g_n))$ is a monomorphism of Z^1 into the n-th direct power M^n of M, on which G acts componentwise. We will in fact calculate $\beta(Z_1)$ and $\beta(B_1)$ using the well-known interpretation of H^1 in a split extension. Let

$$G \rhd\!\!< M = \{(g, m) \mid g \in G, \ m \in M, \ (g_1, m_1)(g_2, m_2) = (g_1 g_2, m_1^{g_2} m_2)\}.$$

Then the set of complements of $M^* = \{(1, m) | m \in M\}$ is described as

$$\mathcal{K} = \{K_\gamma = \{(g, \gamma(g)) \mid g \in G\} \mid \gamma \in Z^1(G, M)\}$$

and two complements K_{γ_1} and K_{γ_2} are conjugate in $G \rhd\!\!< M$ if and only if $\gamma_1 \in B^1(G, M)\gamma_2$ so that there is a bijection between the conjugacy classes of complements and the elements of $H^1(G, M)$.

On the other hand, if

$$1 \to M \overset{id}{\to} H \overset{\pi}{\to} G \to 1$$

is an exact sequence and τ a mapping from $\{g_1, \ldots, g_n\}$ into H with $g_i^{\tau\pi} = g_i$, then this extension splits if and only if there exists a mapping $f : \{g_1, \ldots, g_n\} \to M$ such that for all $j = 1, \ldots, r$

$$R_j(g_1^\tau f(g_1), \ldots, g_n^\tau f(g_n)) = 1, \tag{1}$$

a complement being given by

$$\langle \ g_i^\tau f(g_i) \mid i = 1, \ldots, n \ \rangle.$$

We want to determine these functions f by solving (1) for $(f(g_1), \ldots, f(g_n)) \in M^n$. Since $M \lhd H$, we may in each relation "collect" the elements g_i^τ to the left, without changing their order and we get

$$R_j(g_1^\tau, \ldots, g_n^\tau) \ R_j^*(f(g_1), \ldots, f(g_n)) = 1 \tag{2}$$

where $R_j(g_1^\tau, \ldots, g_n^\tau) \in M$ and where R_j^* is a map from M^n into M with $R_j^*(f(g_1), \ldots, f(g_n))$ a product of conjugates of the elements $f(g_i)$ under the elements g_j^τ. If we write M additively, i.e. as a $\mathbf{Z}_p G$-module of dimension d, conjugation by an element g_j^τ becomes application of a linear map G_j corresponding to g_j, and after introduction of a basis of M becomes right-multiplication of rows by a matrix that we also denote by G_j. R_j^* becomes a linear map from the sum of n copies of this module into M, and hence with respect to suitable bases is represented by a $dn \times d$ matrix. If we express each $f(g_i)$ as a linear combination of the basis with unknown coefficients $x_{i1}, \ldots, x_{id}$, then we see that each relation (2) yields an inhomogeneous system of d equations in the nd unknown x_{ij}, so the r relations yield a system of rd such equations. This fact puts a premium on getting a presentation with a small number of relations. Unfortunately, no general methods seem to be known for producing such presentations.

It should be mentioned, that the "collecting process" producing R_j^* can be described with the help of "Fox derivatives"; in the case of a soluble group G described by a pc presentation the data structure for the implementation is particularly neat.

Any two solutions $(f(g_1), \ldots, f(g_n))$ and $(f'(g_1), \ldots, f'(g_n))$ of this system of equations differ by some $(\gamma_m(g_1), \ldots, \gamma_m(g_n)) \in \beta(Z^1(G, M))$ determined by a cocycle $\gamma_m \in Z^1(G, M)$. I.e., if $L = f_0 + L_{Hom}$ is the solution set of the system of equations, then L_{Hom} describes $\beta(Z^1(G, M))$. In certain cases the number of unknowns in the system of equations for Z^1 can further be reduced. If P' is a p'-subgroup of G then it will split over M, so for generators g_i of G with $g_i \in P'$ we may without loss assume $f(g_i) = 1$.

The computation of $\beta(B^1(G, M))$ follows directly from the definition: for each $\gamma_m \in B^1(G, M)$ there exists $m \in M$ such that $\gamma_m(g) = m m^{-g}$. Using additive notation and identifying M with $\mathbf{Z}_p^{1 \times d}$ via a basis of M again, and using XG_j also to denote the multiplication of the row X by the $d \times d$ matrix corresponding to g_j, we see that if $S := (I - G_1, \ldots, I - G_n)$ is a $d \times dn$ matrix, then $\beta(B^1(G, M))$ is represented by the row space of S. A cocycle γ lies in $B^1(G, M)$ if and only if there exists a solution X of the inhomogeneous linear system $XS = B_\gamma$, where B_γ is the dn-row describing $\beta(\gamma) = (\gamma(g_1), \ldots, \gamma(g_n))$ with respect to the chosen basis of M. Solving this system yields the d-row X describing the element $m \in M$ with $\gamma_m = \gamma$, with respect to the basis. The space of solutions of $XS = 0$ describes $C_M(K)$, where K is an arbitrary complement of M in H. (This space of solutions is of course equal to the space of common eigenvectors of $G_1, \ldots, G_n$ with eigenvalue 1).

The computation of H^1 is demonstrated by a simple example:

Let $S_4 := \langle a, b, c, d \rangle$ with $a = (12)$, $b = (123)$, $c = (13)(24)$, $d = (12)(34)$. Let $H := S_4 \times Z_2 = S_4 \times \langle e \rangle$, let $M := \langle c, d, e \rangle \lhd H$ and $G := \langle g_1, g_2 \mid g_1^2 = g_2^3 = (g_2 g_1)^2 = 1 \rangle$. Then $\pi : a \to g_1, b \to g_2, c \to 1, d \to 1, e \to 1$ is an epimorphism with kernel M, and τ with $g_1^\tau = ac$, $g_2^\tau = b$ fulfills $g_i^{\tau\pi} = g_i$, $i = 1, 2$.
"Collecting" the three relations in H and writing $f_i^{g_j}$ for $(g_j^\tau)^{-1} f(g_i) g_j^\tau$, we get

$$
\begin{aligned}
1 &= (g_1^\tau f(g_1))^2 &&= (g_1^\tau)^2 f_1^{g_1} f_1 &&= d\, f_1^{g_1} f_1, \\
1 &= (g_2^\tau f(g_2))^3 &&= (g_2^\tau)^3 f_2^{g_2^2} f_2^{g_2} f_2 &&= f_2^{g_2^2} f_2^{g_2} f_2, \\
1 &= (g_2^\tau f(g_2) g_1^\tau f(g_1))^2 &&= (g_2^\tau g_1^\tau)^2 f_2^{g_1 g_2 g_1} f_1^{g_2 g_1} f_2^{g_1} f_1 &&= cd f_2^{g_1 g_2 g_1} f_1^{g_2 g_1} f_2^{g_1} f_1.
\end{aligned}
$$

Introducing additive notation and basis vectors $(1\,0\,0) \leftrightarrow c$, $(0\,1\,0) \leftrightarrow d$,

$$
(0\,0\,1) \leftrightarrow e \quad \text{we have } G_1 = \begin{pmatrix} 1 & 1 & 0 \\ 0 & 1 & 0 \\ 0 & 0 & 1 \end{pmatrix}, \; G_2 = \begin{pmatrix} 0 & 1 & 0 \\ 1 & 1 & 0 \\ 0 & 0 & 1 \end{pmatrix}.
$$

With $f(g_1) \leftrightarrow X_1 = (x_{11}, x_{12}, x_{13})$, $f(g_2) \leftrightarrow X_2 = (x_{21}, x_{22}, x_{23})$ we get

$$
(0\,0\,0) = (0\,1\,0) + X_1 G_1 + X_1 = (0\,1\,0) + X_1 \begin{pmatrix} 0 & 1 & 0 \\ 0 & 0 & 0 \\ 0 & 0 & 0 \end{pmatrix},
$$

$$
(0\,0\,0) = X_2 G_2^2 + X_2 G_2 + X_2 = X_2 \begin{pmatrix} 0 & 0 & 0 \\ 0 & 0 & 0 \\ 0 & 0 & 1 \end{pmatrix},
$$

$$
\begin{aligned}
(0\,0\,0) &= (1\,1\,0) + X_2 G_1 G_2 G_1 + X_1 G_2 G_1 + X_2 G_1 + X_1 \\
&= (1\;1\;0) + X_1 \begin{pmatrix} 1 & 1 & 0 \\ 1 & 1 & 0 \\ 0 & 0 & 0 \end{pmatrix} + X_2 \begin{pmatrix} 0 & 0 & 0 \\ 1 & 1 & 0 \\ 0 & 0 & 0 \end{pmatrix}.
\end{aligned}
$$

So with $X = (x_{11}, x_{12}, x_{13}, x_{21}, x_{22}, x_{23})$ and $V = (0\,1\,0 \quad 0\,0\,0 \quad 1\,1\,0)$

$$X \begin{pmatrix} 0 & 1 & 0 & 0 & 0 & 0 & 1 & 1 & 0 \\ 0 & 0 & 0 & 0 & 0 & 0 & 1 & 1 & 0 \\ 0 & 0 & 0 & 0 & 0 & 0 & 0 & 0 & 0 \\ 0 & 0 & 0 & 0 & 0 & 0 & 0 & 0 & 0 \\ 0 & 0 & 0 & 0 & 0 & 0 & 1 & 1 & 0 \\ 0 & 0 & 0 & 0 & 0 & 1 & 0 & 0 & 0 \end{pmatrix} = V.$$

The set of solutions X is

$$(1\,0\,0 \quad 0\,0\,0) + \left\langle \begin{matrix} (0 & 1 & 0 & 0 & 1 & 0) \\ (0 & 0 & 1 & 0 & 0 & 0) \\ (0 & 0 & 0 & 1 & 0 & 0) \end{matrix} \right\rangle.$$

$B^1(G, M)$ is determined by the rowspace of

$$S = (G_1 - I, G_2 - I) = \begin{pmatrix} 0 & 1 & 0 & 1 & 1 & 0 \\ 0 & 0 & 0 & 1 & 0 & 0 \\ 0 & 0 & 0 & 0 & 0 & 0 \end{pmatrix}.$$

So $H^1(G, M)$ is isomorphic to $(0\,0\,1 \quad 0\,0\,0) + B^1(G, M)$ and M has two classes of complements, represented by $\langle a, b\rangle$ and $\langle ae, b\rangle$.

4. Complements.

We are next going to describe a general method for finding the conjugacy classes of complements of an arbitrary normal subgroup M of a finite soluble group G, using a similar kind of homomorphism principle as for the classes of elements.

If M is elementary abelian, we can of course compute $H^1(G, M)$ as just described. Otherwise, using the non-commutative Gauß algorithm, it is not difficult to construct a series of normal subgroups with elementary abelian factors passing through M, that is:

$$G = N_1 > N_2 > \cdots > N_i = M > \cdots > N_m = \langle 1 \rangle$$

with $N_j \lhd G$ and N_j/N_{j+1} elementary abelian for $j = 1, \ldots, m-1$. Starting with $j = i$ and stepping down for $j = i+1, \ldots, m$, we will construct inductively the classes of complements of M/N_j in G/N_j. It suffices to describe a typical step from G/N_j to G/N_{j+1}. I.e., we may assume that

$N = N_j$ is elementary abelian and that the results are known for G/N_j. We shall need the following well-known

4.1 Lemma: Let G be a group, let $M \lhd G$ and let $K_1, K_2 < G$ be complements of M in G. Then K_1 is conjugate to K_2 in G if and only if K_1 is conjugate to K_2 under M. Furthermore $N_M(K) = C_M(K)$ for each complement K of M and $N_G(K) = K\,C_M(K)$.

From this we get a reduction scheme, the proof of which is immediate from the figure.

4.2 Theorem: Let G be a group with $M, N \lhd G$ and $M > N$.

(i) Let $\{U_1/N, \ \dots \ , U_k/N\}$ be a system of representatives of the conjugacy classes of complements of M/N in G/N.

(ii) For each $i = 1, \ \dots \ , k$ let $\overline{C}_i = C_{M/N}(U_i/N)$ and let C_i be its complete preimage in G, that is, $C_i/N = \overline{C}_i$.

(iii) Let $\mathcal{K}_i$ be the set of complements of N in U_i, (the empty set, if none exist), let $\mathcal{L}_i = \{K_{i1}, \ \dots, \ K_{is_i}\}$ be a set of representatives of the orbits of C_i acting on $\mathcal{K}_i$ by conjugation (or again the empty set), and let $C_{ij} = C_{C_i}(K_{ij})$.

Then $\bigcup_{i=1}^{k} \mathcal{L}_i$ is a set of representatives of the conjugacy classes of complements of M in G, and $C_{ij} = C_M(K_{ij})$.

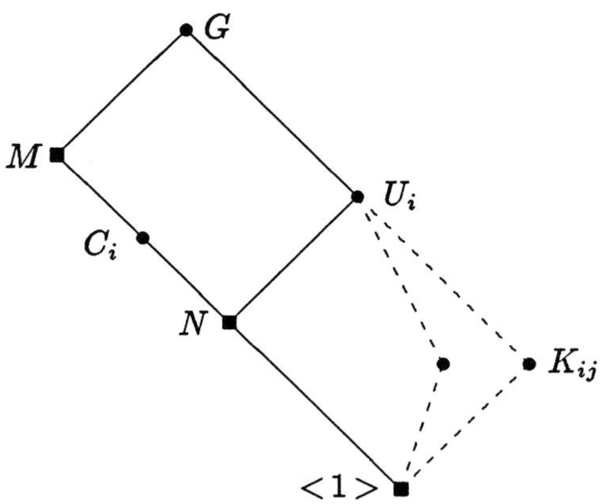

To apply Theorem 4.2 inductively in the above mentioned "typical step", we have to perform the following tasks for each U_i:

(a) We have to decide whether N has a complement in U_i. Since N is elementary abelian in our situation, one way this can be done is by investigating the solvability of an inhomogeneous system of linear equations, as explained in section 3.

(b) Let us then assume that there is a complement K of N in U_i. Then each $K_{ij} \in \mathcal{K}_i$ can be written as

$$K_{ij} = \{\ k\gamma_j(k)\ \mid\ k \in K\}$$

with a cocycle $\gamma_j \in Z^1(K,N) \cong Z^1(U_i/N,N)$. The group C_i operates on $\mathcal{K}_i$ by conjugation. Let $c \in C_i$, then

$$c^{-1}K_{ij}c = \ \{\ c^{-1}k\gamma_j(k)c\ \mid\ k \in K\} = \ \{\ k \cdot [k,c] \cdot c^{-1}\gamma_j(k)c\ \mid\ k \in K\},$$

and since $c^{-1}K_{ij}c$ is also a complement of N in U_i, we see that $\alpha_c(\gamma_j)$ defined by

$$(\alpha_c(\gamma_j))(k)\ := \ [k,c]c^{-1}\gamma_j(k)c\ \in N$$

is also a cocycle in $Z^1(K,N)$.

Since $cN\ \in C_{G/N}(U_i/N)$, we see that the element $[k,c]$ of N is independent of γ_j, while $\gamma_j(k)\ \rightarrow\ c^{-1}\gamma_j(k)c$ defines a linear mapping of $Z^1(K,N)$ again considered as a $\mathbf{Z}_p$-vector space. Hence

$$\alpha_c\ :\ \gamma_j\ \rightarrow\ \alpha_c(\gamma_j)$$

is an affine mapping of the $\mathbf{Z}_p$-vector space $Z^1(K,N)$ onto itself. Now let $n \in N\ \lhd\ C_i$. Then

$$(\alpha_n(\gamma_j))(k) = [k,n]n^{-1}\gamma_j(k)n = [k,n]\gamma_j(k).$$

But $k \rightarrow [k,n] = n\,n^{-k}$ is just the coboundary γ_n, that is, $(\alpha_n(\gamma_j))(k) = \gamma_j(k)\,\gamma_n(k)$. Hence $N\ \lhd\ C_i$ acts on $Z^1(K,N)$ as the group of "translations" by the coboundaries, and the orbits of N on Z^1 are the cosets of B^1. Therefore C_i/N and hence C_i act on $H^1(K,N) = Z^1(K,N) \,/\, B^1(K,N)$ again as a group of affine mappings

$$\overline{\alpha}_c = \overline{\alpha}_{cN} \;:\; H^1(K,N) \;\to\; H^1(K,N)$$

with

$$\overline{\alpha}_{cN}(\gamma_j\, B^1(K,N)) = \alpha_c(\gamma_j)\, B^1(K,N).$$

(c) It remains to find $C_{ij} := C_{C_i}(K_{ij})$ for each $K_{ij} \in \mathcal{L}_i$, the set of representatives of the C_i-orbits on $\mathcal{K}_i$. $C_{C_i}(K_{ij})$ can be described as the stabilizer of the cocycle γ_j belonging to K_{ij} in the affine action via α of C_i on $Z^1(K,N)$. The orbit-stabilizer algorithm applied to the action via $\overline{\alpha}$ of C_i on $H^1(K,N)$ will give the stabilizer $\overline{C}_{ij}$ of $\gamma_j B^1(K,N) \in H^1$. In terms of the action of C_i via α on $Z^1(K,N)$, of course, $\overline{C}_{ij}$ is the block stabilizer of the block $\gamma_j\, B^1(K,N)$ which was obtained as the orbit of γ_j under the action of N. Hence for each $c \in \overline{C}_{ij}$ there exists $n_c \in N$ such that

$$\alpha_c(\gamma_j) = \alpha_{n_c}(\gamma_j) = \gamma_j\, \gamma_{n_c}.$$

Therefore the values of the coboundary γ_{n_c} are determined as

$$\gamma_{n_c}(k) = (\gamma_j(k))^{-1}\, (\alpha_c(\gamma_j))(k).$$

The element n_c can be determined by a solution of an inhomogeneous system of linear equations and the space of solutions of the homogeneous system determines $C_N(K_{ij})$ as explained before the example in section 3. This allows to compute C_{ij} using the homomorphism principle described at the end of section 2 by solving such an inhomogeneous system of linear equations for each $c \in \{c_1, \ldots, c_k\}$ with $\langle c_1 N, \ldots, c_k N\rangle = \overline{C}_{ij}/N$.

5. Special cases and further reductions.

5.1 The central case.

The method described in section 4 reduces the use of the rather time-consuming orbit-stabilizer algorithm to the action of C_i on $H^1(K,N)$; the rest is all done by solving systems of linear equations. So the reduction is best if $H^1(K,N)$ is small, i.e., if B^1 is close to Z^1. On the other hand, if N is a central subgroup, then B^1 is trivial, and hence no reduction of this kind is possible. However, in this case linear methods are again possible, in a way analogous to those described in [MeN 89] (see also [LNS 84]).

Using the notation of the last section, let us again assume that there exists a complement K of N in U_i. The complete preimage of $N_{G/N}(U_i/N)$ is equal to C_iU_i by lemma 4.1. Let N be central in C_iU_i. Since $K < C_iU_i$ centralizes N, we have $B^1(K,N) = 1$ and $H^1(K,N) = Z^1(K,N)$. For $\gamma \in Z^1(K,N)$ and $c \in C_i$

$$(\alpha_c(\gamma))(k) = [k,c]c^{-1}\gamma(k)c = [k,c]\gamma(k) \quad \text{for all } k \in K.$$

Applying this equation to the trivial cocycle γ_0, defined by $\gamma_0(k) = 1$ for all $k \in K$, we see that each $c \in C_i$ defines a cocycle $\gamma_c : k \to [k,c]$. Because $[k,c][k,c'] = [k,cc']$ these form a subgroup S_i of $Z^1(K,N)$ and the orbits of C_i on $Z^1(K,N)$ are the cosets of this subgroup, so can easily be described. It remains to find $C_{C_i}(K_{ij})$ for each K_{ij}. Since N is centralized by C_i, all these centralizers are equal: $C_{C_i}(K_{ij}) = C_{C_i}(K)$ for all j and this subgroup can be found by the following lemma which is an exact analogue of 3.1 in [MeN 89]:

Lemma: Let $\{g_1,\ldots,g_m,g_{m+1},\ldots,g_n\}$ be a PAG sequence for C_i, with $\langle g_{m+1},\ldots,g_n\rangle = N$. Then $\langle \gamma_1 = \gamma_{g_1},\ldots,\gamma_m = \gamma_{g_m}\rangle = S_i < Z^1(K,M)$. Adopting additive notation, choose a basis $\{\gamma_{i_s},\ldots,\gamma_{i_1} \mid m \geq i_s > \ldots > i_1 \geq 1\}$ for S_i running from γ_m to γ_1 in this order. Then for each $k \in \{1,\ldots,m\}\backslash\{i_1,\ldots,i_s\}$ we can express γ_k as

$$\gamma_k = \gamma_{i_1}^{e_{k,i_1}} \cdots \gamma_{i_s}^{e_{k,i_s}},$$

and $C_{C_i}(K)$ has the PAG sequence

$$\{g_k(g_{i_1}^{e_{k,i_1}} \cdots g_{i_s}^{e_{k,i_s}})^{-1} \mid k \in \{1,\ldots,m\}\backslash\{i_1,\ldots,i_s\}\} \cup \{g_{m+1},\ldots,g_n\}.$$

In practical implementation, of course, a PAG sequence of G is passed through N and a CGS is used for each subgroup that occurs.

5.2 Normal complements.

If one only wants to get normal complements one is able to treat an even more general situation. Let G be a soluble normal subgroup of the group H and let $M \lhd H$ with $M < G$. Then we can obtain all $K \lhd H$ with $MK = G$ and $M \cap K = \langle 1 \rangle$ by inductive use of the following

Lemma: Let H be a group and $G, M, N \lhd H$ with $G > M > N$. Let N be elementary abelian and $U \lhd H$ with $MU = G$ and $M \cap U = N$.

Let $U/N = \langle u_1N, \ldots, u_nN \mid R_j(u_1N, \ldots, u_nN) = 1,\ j = 1, \ldots, r \rangle$,

let $M = \langle m_1, \ldots, m_s \rangle$, and

let $H/G = \langle h_1G, \ldots, h_tG \rangle$ and $(u_iN)^{h_lN} = W_{il}(u_1N, \ldots, u_nN)$ for $i = 1, \ldots, n$ and $l = 1, \ldots, t$.

Then each function $f : \{u_1, \ldots, u_n\} \to N$, satisfying

(i) $R_j(u_1f(u_1), \ldots, u_nf(u_n)) = 1$ for $j = 1, \ldots, r$,

(ii) $[m_k, u_if(u_i)] = 1$ for $k = 1, \ldots, s$ and $i = 1, \ldots, n$, and

(iii) $(u_if(u_i))^{h_l} = W_{il}(u_1f(u_1), \ldots, u_nf(u_n))$ for $i = 1, \ldots, n$ and $l = 1, \ldots, t$

determines a subgroup $V \lhd H$ with $MV = G$ and $M \cap V = \langle 1 \rangle$ as $V = \langle u_1f(u_1), \ldots, u_nf(u_n) \rangle$, and each $V \lhd H$ with $MV = G$ and $M \cap V = \langle 1 \rangle$ is determined by such an f in this way.

The proof uses the fact that V will centralize M, and the computational use of conditions (i)-(iii) for the determination of all complements V is analogous to the procedure described in section 4.

5.3 Further reductions.

In this section we give a brief sketch of some reduction methods for finding complements. Consider again the general situation described in section 4. We have $M \lhd G$ and want complements to M in G. Assuming that we have $N \lhd G$ with $N \lhd M$ and we have found a complement U/N to M/N in G/N, we aim to choose $N_1 \lhd G$ with N/N_1 elementary abelian, and then find complements to N/N_1 in U/N_1. For simplicity of exposition, we will not pay attention here to the management of conjugacy classes; the basic tool for that is Lemma 4.1. In the notation of section 4, everything takes place inside C_iU_i, so we may as well assume here that $C_i = M$, $U_i = U \lhd G = MU$ and $M \cap U = N > N_1 \geq \langle 1 \rangle$.

The basic idea is to reduce the problem of finding complements to N/N_1 in U/N_1 to a problem for smaller groups. If we limit our computations to the linear algebra methods of section 3, the number of linear equations from a PAG presentation of U/N (i.e., of G/M) grows quadratically with the composition length of U/N, and the number of coefficients in the equations grows as the cube of the composition length. For moderately large groups, it may be sensible or necessary to cut down the size before computing. Of course there are many groups, an easy example is the quaternion group, in which the number of classes of complements is bigger in some factor groups than it is in the whole group, and it is not always clear a priori that reduction in size is not buying too many classes in exchange. This problem is, in fact, already present for the general scheme described in section 4.

One way to reduce the size of G/N_1 is to replace N_1 by $N_2 \lhd G$ with $N_1 < N_2 < N$ and to consider G/N_2 first. The idea is to choose N_1 dynamically as the computation proceeds and to be willing to change the choice to N_2 in the light of experience. We will shortly give an example of this sort of replacement.

Another way to reduce G/N_1 is to replace N_1 by $L \lhd G$ with $N \cap L = N_1$ and $L \lhd U$, and then to complement NL/L in U/L. This device is particularly effective if there is a subgroup L, necessarily normal in G, that is contained in every complement to N/N_1 in U/N_1.

Here is an illustration, with $N_1 = \langle 1 \rangle$ for convenience and with N an elementary abelian p-group. Let $C_U(N) =: D$ and let $[D, D]D^p =: E$. Then E is in every complement to N in U. If $\langle 1 \rangle < E$, then we can recur to G/E to find complements to NE/E in U/E. If $E = \langle 1 \rangle$, then D is itself a $\mathbf{Z}_p$-vector space on which U acts. In this case, complements to N in U intersect D in $\mathbf{Z}_pU$-submodules W that complement N in D. One can, in effect, recur to U/W unless $W = \langle 1 \rangle$, i.e., unless $N = C_U(N)$.

Suppose that $N = C_U(N)$. Then we can perhaps use Sylow theory as well as linear algebra. If $N = U$, then of course we are done. Otherwise, suppose that $N < Q \lhd U$ with Q/N elementary abelian, which is more than we need, but easy to produce. If Q is also a p-group, then $1 < [N, Q]$

because $N = C_U(N)$, so we can recur to $G/[N,Q]$ to get subgroups Y that complement $N/[N,Q]$ in $U/[N,Q]$. Then replace N by $[N,Q]$ and recur to get complements to $[N,Q]$ in each Y. If, on the other hand, Q/N is a p'-group, then the complements to N in Q are the (conjugate) p-complements R of Q. By the Frattini argument, $U = N_U(R)N$, and $N \cap N_U(R) = C_N(R) = C_N(Q) \lhd G$. If $C_N(Q) \neq \langle 1 \rangle$, we can recur to $G/C_N(Q)$, while if $C_N(Q) = \langle 1 \rangle$, the subgroup $N_U(R)$ is a representative complement to N in U.

This account gives only a hint of some of the kinds of reductions that are possible and the tools that they require. Plainly, this sort of attack would only make sense in a setting in which the size of G/M made solving systems of linear equations unattractive or impractical. The general design of recursive algorithms for finding complements is a fairly complex question that we must leave for another paper.

6. Normalizers.

The ideas described in the previous sections can also be used to supplement a method for the calculation of the normalizer $N_G(U)$ of an arbitrary subgroup of a soluble group, which has been proposed and implemented in CAYLEY by S. Glasby and M. Slattery [GlS 90]. While their method works very efficiently in many cases, there are others in which they have to resort to explicit calculation of large orbits of subgroups under conjugation. It is in these cases that we can offer improvement. The case distinction will become clear in the sequel and examples will be given in section 7.

As with the tasks discussed in the last sections, the computation of $N_G(U)$ can be split up inductively by virtue of the following easily proved facts.

6.1 Lemma: Let H be a group, $V < H$, $N \lhd H$ and $N < S \leq H$ such that $\bar{S} := S/N = N_{H/N}(VN/N)$. Then $N_H(V) \leq S$ and hence $N_H(V) = N_S(V)$.

6.2 Lemma: Let H be a group, $V < H$, $M \lhd H$ and $T = N_H(V \cap M)$. Then $N_H(V) \leq T$ and hence $N_H(V) = N_T(V)$.

Glasby and Slattery combine these two relations in the following way: Let

$$G = N_1 \rhd N_2 \rhd \ldots \rhd N_m = \langle 1 \rangle$$

be a series of normal subgroups $N_i \lhd G$. For $1 \leq j < i \leq m$, that is, for $N_i < N_j$, let

$$U_{ij} := (UN_i \cap N_j) = (U \cap N_j)N_i \quad \text{and} \quad \overline{U}_{ij} := U_{ij}/N_i.$$

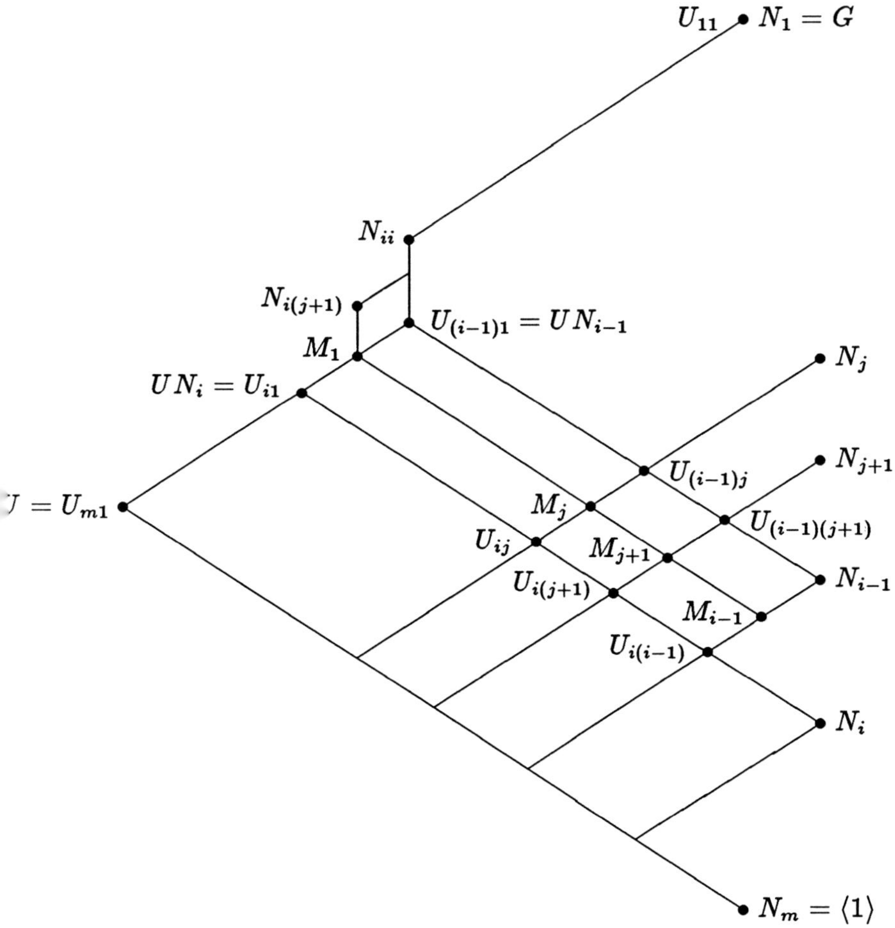

Define $N_{11} := G$, $\overline{N}_{11} := G/N_1 = \langle 1 \rangle$ and by induction for each $i = 2, \ldots, m$:

(for $j = i$)	let N_{ii} be the complete preimage in G of $\overline{N}_{(i-1)1} = N_{G/N_{i-1}}(\overline{U}_{(i-1)1})$ and $\overline{N}_{ii} := N_{ii}/N_i$
for $j = i-1, \ldots, 1$	let $\overline{N}_{ij} := N_{\overline{N}_{i(j+1)}}(\overline{U}_{ij})$ and N_{ij} be the complete preimage of $\overline{N}_{ij}$ in G.

Then $N_G(U)$ is obtained as N_{m1}. The statement is seen by twofold induction:

For fixed i by definition $\overline{N}_{i(i-1)} = N_{\overline{N}_{ii}}(\overline{U}_{i(i-1)})$, so (by lemma 6.2 with $H := \overline{N}_{ii}$) we have $\overline{N}_{ij} = N_{\overline{N}_{ii}}(\overline{U}_{ij})$ for $j = i-2, \ldots, 1$, and hence in particular $\overline{N}_{i1} = N_{\overline{N}_{ii}}(\overline{U}_{i1})$. By induction on i, using lemma 6.1, this proves the claim.

Computationally there are two different kinds of steps in this induction:

1. $\overline{N}_{i(i-1)} := N_{\overline{N}_{ii}}(\overline{U}_{i(i-1)})$ has indeed to be found by an "orbit-stabilizer" algorithm (cf. [LNS 84], 3.1, 3.2, p. 111). However one can take advantage of the fact that $\overline{U}_{i(i-1)} = (UN_i \cap N_{i-1})/N_i \leq N_{i-1}/N_i$ which is an elementary abelian normal subgroup of $\overline{N}_{ii} = N_{ii}/N_i$, and hence can be considered as a $\mathbf{Z}_p$-vector space. The matrices describing the action of the generators of $\overline{N}_{ii}$ on N_{i-1}/N_i can be computed once and for all, the non-commutative Gauß algorithm for the images of $\overline{U}_{i(i-1)}$ reduces to an ordinary one and only vector addition and matrix multiplication are needed instead of the much more time-consuming collection algorithm.

2. On the other hand, for the determination of $\overline{N}_{ij} := N_{\overline{N}_{i(j+1)}}(\overline{U}_{ij})$ with $j = i-2, \ldots, 1$ we can use the methods developed in the previous sections.

   Since by definition $U_{i(j+1)} \lhd N_{i(j+1)}$ and $U_{i(j+1)} \leq U_{ij} \leq N_{ij} \leq N_{i(j+1)}$, we have

   $$\overline{N}_{ij}/\overline{U}_{i(j+1)} \cong N_{ij}/U_{i(j+1)} = N_{N_{i(j+1)}/U_{i(j+1)}}(U_{ij}/U_{i(j+1)}).$$

   To determine this group we introduce $M_1 := N_{i(j+1)} \cap U_{(i-1)1}$ and $M_k := N_{i(j+1)} \cap U_{(i-1)k} = M_1 \cap U_{(i-1)k}$ for $k = 2, \ldots, i-1$. As we have seen, $U_{i1} \leq N_{i1} \leq N_{i(j+1)} \leq N_{ii}$ and $U_{i1} \leq U_{(i-1)1} \lhd N_{ii}$. Therefore

$U_{(i-1)1}$ and further all $U_{(i-1)k} = U_{(i-1)1} \cap N_k$, $k = 2, \dots, i-1$ and all M_k, $k = 1, \dots, i-1$ are normalized by $N_{i(j+1)}$. Finally, since

$$U_{(i-1)k} = U_{ik} U_{(i-1)l}, \quad \text{and}$$

$$U_{il} = U_{ik} \cap U_{(i-1)l}$$

for $k = 1, \dots, i-1$ and $l = k+1, \dots, i-1$, we also get

$$M_k = U_{ik} M_l \quad \text{and}$$

$$U_{il} = U_{ik} \cap M_l$$

for $k = 1, \dots, i-1$ and $l = k+1, \dots, i-1$. From this we see that if for some j we have $U_{ij} = M_j \lhd N_{i(j+1)}$ then for all $k < j$ also $U_{ik} = M_k$, in particular $U_{i1} = M_1 \lhd N_{i(j+1)}$ and hence $N_{i1} = N_{i(j+1)}$, so that we can immediately proceed from G/N_i to G/N_{i+1}.

Otherwise we have

$$U_{ij} M_{j+1} = M_j \lhd N_{i(j+1)} \quad \text{and}$$

$$U_{ij} \cap M_{j+1} = U_{i(j+1)} \lhd N_{i(j+1)} \quad .$$

That is, $M_j/U_{i(j+1)}$ is a semidirect product of the elementary abelian normal subgroup $M_{j+1}/U_{i(j+1)}$ by $U_{ij}/U_{i(j+1)}$ and since M_j and M_{j+1} are normal in $N_{i(j+1)}$ we have that $N_{N_{i(j+1)}/U_{i(j+1)}}(U_{ij}/U_{i(j+1)})$ is the stabilizer of $U_{ij}/U_{i(j+1)}$ in the operation of $N_{i(j+1)}/U_{i(j+1)}$ on the set of complements. The orbits of M_{j+1}, which are also orbits of M_j, are blocks in this operation, and $N_{i(j+1)} \,/\, U_{i(j+1)}$, or equivalently $N_{i(j+1)} \,/\, M_j$, operates on these blocks, which can be described by the elements of $H^1(U_{ij} \,/\, U_{i(j+1)}, M_{j+1} \,/\, U_{i(j+1)})$.

We have to find the stabilizer of the block containing U_{ij}, that is the stabilizer of the element $B^1(U_{ij} \,/\, U_{i(j+1)}, M_{j+1} \,/\, U_{i(j+1)}) \in H^1$ by the orbit-stabilizer method, but we can then get from this the desired $N_{ij} \,/\, U_{i(j+1)}$ as the stabilizer of the trivial cocycle in the same way as described at the end of section 4(c).

The group $M_{j+1}/U_{i(j+1)}$ is isomorphic to a factor of N_{i-1}/N_i which is an elementary abelian p-group, and $U_{ij}/U_{i(j+1)}$ is isomorphic to a subgroup of N_j/N_{j+1} which is an elementary abelian q-group.

If $p \neq q$, $H^1(U_{ij}/U_{i(j+1)}, M_{j+1}/U_{i(j+1)}) = 1$, so there is only one block of complements and $N_{i(j+1)}/U_{i(j+1)}$ is its stabilizer, no orbit-stabilizer algorithm is necessary, and only the task to solve the inhomogeneous sets of linear equations remains in order to determine $N_{ij}/U_{i(j+1)}$. For each inhomogeneous set of equations one solution can be written down directly as described in [GlS 90].

7. Performance.

Comparison with other implementations is only possible for the normalizer routine for soluble groups in CAYLEY described in [GlS 90]. As is to be expected, a gain is reached if orbit-stabilizer calculations can be replaced by "linear" methods in the "non-coprime" situation, as in the following examples suggested by M. Slattery.

Let p and q be primes, $n \in \mathbf{N}$ and $p \mid n \mid q-1$. Let Z_p, Z_q and $Z_n = \langle a \rangle$ be cyclic, $Z_q \rtimes Z_n$ be a nonabelian semidirect product represented as a permutation group of degree q and $G = Z_p\, wr_q\, (Z_q \rtimes Z_n)$. We list for some values of p, q and n and subgroups $U < \langle a \rangle$ the order of $N_G(U)$ and as indication of the length of the longest orbit occurring in an orbit-stabilizer method the order of some B^1, and we give computing times for the determination of $N_G(U)$: t_1 for the GAP-implementation on a MASSCOMP 5500 PEP, t_2 for the CAYLEY implementation on an APOLLO 10000, which is 5-10 times faster.

p	q	n	$\lvert G\rvert$	U	$\lvert N(U)\rvert$	$\lvert B^1\rvert$	t_1	t_2
2	31	30	$2^{32} \cdot 3 \cdot 5 \cdot 31$	$\langle a^{15} \rangle$	$2^{17} \cdot 3 \cdot 5$	2^{15}	25 *sec*	103 *sec*
3	13	12	$3^{14} \cdot 2^2 \cdot 15$	$\langle a^4 \rangle$	$2^2 \cdot 3^6$	3^8	10 *sec*	10 *sec*
5	11	10	$5^{12} \cdot 2 \cdot 11$	$\langle a^2 \rangle$	5^4	5^8	6 *sec*	$>$ 400 *sec*

We have further compared implementations, both in GAP, of the Glasby-Slattery method of the "conjugating element" to which we referred at the end of section 6, i.e. without orbit-stabilizer computation, with our method of solving inhomogeneous linear equations instead in the case of a certain group of order $5^6 \cdot 313^5$ and have obtained computing times of 104 *sec* and 20 *sec* respectively, indicating that with elements of higher order the transition to linear methods pays, even in comparison to highly efficient methods involving collection.

8. Concluding remarks

The only parts of the methods described in this paper that have not been implemented yet are the further reductions described in section 5.3. They are actually only part of a more general recursive scheme proposed by the third author in [Wri 88a,b].

It should further be remarked that not only is there a formal analogy between our methods and those of [MeN 89], the latter can in fact be interpreted as using H^0 instead of H^1. The analogy raises the question of the possibility of using similar methods with H^2 for the classification of soluble groups; Zassenhaus' space groups algorithm should provide a model.

Bibliography

[Can 84] Cannon, J., *A Computational Toolkit for Finite Permutation Groups.* M. Aschbacher et al., ed., Proc. Rutgers Group Theory Year 1983/84, Cambridge U.P. 1984, 1–18.

[Con 90] Conlon, S.B., *Calculating characters of p-groups.* J. Symbolic Comp. to appear.

[FeN 79] Felsch, V., Neubüser, J., *An algorithm for the computation of conjugacy classes and centralizers in p-groups.* Lecture Notes in Computer Science 72 (1979) 452–465.

[GlS 90] Glasby, S.P., Slattery, M.C., *Computing intersections and normalizers in soluble groups.* J. Symbolic Comp. to appear.

[HaN 76] Havas, G., Nicholson, T., *Collection.* R.D. Jenks, ed., SYMSAC'76, Assoc. Comp. Mach., New York (1976) 9–14.

[Hol 84] Holt, D.F., *The calculation of the Schur multiplier of a permutation group.* M.D. Atkinson, ed., Computational Group Theory, Acad. Press (1984) 307–319.

[L-GS 90] Leedham-Green, C.R., Soicher, L.H., *Collection from the Left and other Strategies.* J. Symbolic Comp. to appear.

[LNS 84] Laue, R., Neubüser, J., Schoenwaelder, U., *Algorithms for finite soluble groups and the SOGOS system.* M.D. Atkinson, ed., Computational Group Theory, Acad. Press (1984) 105–135.

[Mac 74] Macdonald, I. D., *A computer application to finite p-groups.* J. Austral. Math. Soc. 17 (1974) 102–112.

[MeN 89] Mecky, M., Neubüser, J., *Some remarks on the computation of conjugacy classes of soluble groups.* Bull. Austral. Math. Soc. 40 (1989) 281–292.

[Neu 61] Neubüser, J., *Bestimmung der Untergruppenverbände endlicher p-Gruppen auf einer programmgesteuerten elektronischen Dualmaschine.* Numer. Math. 3 (1961) 271–278.

[New 76] Newman, M.F., *Calculating presentations for certain kinds of quotient groups.* R.D. Jenks, ed., SYMSAC'76, Assoc. Comput. Mach., New York (1976) 2–8.

[NNS 88] Niemeyer, A., Nickel, W., Schönert, M., *GAP. Getting started and Reference Manual.* Lehrstuhl D für Mathematik, RWTH Aachen, 1988.

[O'Br 90] O'Brien, E.A., *The p-group generation algorithm.* J. Symbolic Comp. to appear.

[Ple 87] Plesken, W., *Towards a soluble quotient algorithm.* J. Symbolic Comp. 4 (1987) 111–122.

[Sim 87] Sims, C.C., *Verifying Nilpotence.* J. Symbolic Comp. 3 (1987) 231–247.

[Sim 90a] Sims, C.C., *Implementing the Baumslag-Cannonito-Miller Polycyclic Quotient Algorithm.* J. Symbolic Comp. to appear.

[Sim 90b] Sims, C.C., *Computing the Order of a Solvable Permutation Group.* J. Symbolic Comp. to appear.

[Va-L 90] Vaughan-Lee, M.R., *Collection from the Left.* J. Symbolic Comp. to appear.

[Wri 88a] Wright, C.R.B., *A Pseudo-Cayley complementation procedure for soluble groups.* Manuscript 1988, University of Oregon.

[Wri 88b] Wright, C.R.B., *Recursive Algorithms for computing complements in Finite Groups.* Manuscript 1988, University of Oregon.

(Received: October 1990)

Acta Applicandae Mathematicae **21**: 77–103, 1990.

Methods for computing in algebraic geometry and commutative algebra Rome, March 1990

Mike Stillman *
Department of Mathematics, Cornell University, U.S.A.

March 16, 1990

1 Lecture # 1: Basic methods for computing in commutative algebra and algebraic geometry

Computation in commutative algebra and algebraic geometry has taken a major step forward in the last decade. This is due to the development of Gröbner bases, computer algebra systems which are readily available to researchers, and the great increase in power of computers in this last decade.

In these two lectures we survey some of the computations which are now feasible in commutative algebra and algebraic geometry. The first talk is mainly an overview. We present a handful of useful computations which are all based on Gröbner bases. These are the building blocks for computing more difficult objects. Most of this material is folklore in the subject, although some represents joint work with either D. Bayer or D. Eisenbud.

In the second talk, we consider a specific problem which is important in both commutative algebra and algebraic geoemtry: Given an ideal in a polynomial ring, find its *radical.* The method we use is due to David Eisenbud and Craig Huneke ([6]). Solving this problem involves using the basic building blocks from the first lecture.

*Partially supported by MSI, and NSF.

AMS subject classification (1980). 13Axx.
Key words. Gröbner bases, radical of ideals.

The first question is: what does one want to be able to compute in commutative algebra and algebraic geometry. The following is a partial list:

Operations on rings: Find the kernel of a ring map, the integral closure of a domain, the blowup ring, and the normal cone (associated graded ring).

Operations on ideals: Find: the intersection of a set of ideals, the ideal quotient of two ideals, the radical of an ideal, the asociated primes of an ideal, the primary decomposition of an ideal, the Hilbert function of a homogeneous ideal, the codimension (or dimension) of an ideal.

Operations on ideals and modules: Find the finite free resolution of a module, the annihilator of a module, the module of homomorphisms $\mathrm{Hom}(M, N)$, the dual and double dual of a module, the homology of a complex: $M' \rightarrow M \rightarrow M''$, Ext and Tor modules.

Operations in algebraic geometry: Find the module corresponding to the normal sheaf, canonical sheaf, tangent sheaf, the image of a map corresponding to some sections of a given line bundle, the cohomology of a sheaf on projective space, secant loci, tangent developable, singular locus.

Operations in local algebra: Find the tangent cone of a local ring, the inverse system of a zero-dimensional ideal, a minimal set of generators for an ideal.

The list is quite extensive. There are large areas not mentioned in this list, such as Chow groups and deformations. Currently, most everything on the above list can be computed, and most of these operations have been implemented in Macaulay, in a joint project with D. Eisenbud. The major exceptions are integral closures and primary decompositions. Although algorithms exist, they have not yet been completely implemented. We plan to implement these in Macaulay in the future.

2 Gröbner bases and syzygies

Every computation we will discuss is based on the construction of Gröbner bases. The algorithm for finding a Gröbner basis is originally due to Buchberger in his 1965 thesis ([4]). In 1978, Spear and Schreyer independently ([15], [13]) observed that one could also compute a basis for the *syzygy* module of an ideal (defined later in this section). As far as applications in commutative algebra and algebraic geometry are concerned, this was an important step. Previously, no complete algorithms (other than Hilbert's original method of 1890!) were known.

In this section we review Gröbner bases.

Given an ideal $I = (f_1, \ldots, f_r) \subset S = k[x_1, \ldots, x_n]$, we want to find a better basis for I. The construction of this basis should generalize row reduction of matrices (in the case that each f_i is linear), and the Euclidean algorithm for finding the g.c.d. of polynomials (in the case that $S = k[x]$ has only one variable.)

In order to generalize these constructions to the non-linear multi-variable case, we order the monomials of S:

Definition 1 A *multiplicative order* (sometimes also called an *admissible* order) is a total order, $>$, on the monomials of S satisfying the following two conditions.

(a) $x^A > 1$, for every monomial $x^A \neq 1 \in S$, and

(b) If $x^A > x^B$, then for every monomial $x^C \in S$, $x^A x^C > x^B x^C$.

Definition 2 The *graded lexicographic order* is the order $>$ defined by $x^A > x^B$ if and only if either $\deg x^A > \deg x^B$ or $\deg x^A = \deg x^B$ and the first non-zero component of the vector $A - B$ is positive.

The *graded reverse lexicographic order* is the order $>$ defined by $x^A > x^B$ if and only if either $\deg x^A > \deg x^B$ or $\deg x^A = \deg x^B$ and the last non-zero component of the vector $A - B$ is negative.

Example 3 If $S = k[x, y, z]$, and $>$ is the graded lexicographic order, then

$$x^2 > xy > xz > y^2 > yz > z^2,$$

whereas if $>$ is the graded reverse lexicographic order, then

$$x^2 > xy > y^2 > xz > yz > z^2.$$

These two orders are *not* the same up to permutation of the variables. They have quite distinct properties, as we will see later.

There are an infinite number of different (multiplicative) orders on S.

Definition 4 Given a non-zero polynomial $f \in S$, and a multiplicative order $>$, define the *initial term* of f, denoted $\mathrm{in}_>(f)$, or more simply as just in(f), to be the greatest monomial of f in the order $>$. For example, if $>$ is the graded reverse lexicographic order, then $\text{in}(2xy + 3y^2 - xz) = xy$.

Given an ideal $I \subset S$, define the *initial ideal* of I, denoted by $\text{in}_>(I)$, or $\text{in}(I)$ to be the ideal generated by the monomials $\{\text{in}(f) \mid f \in I\}$.

Since S is a Noetherian ring, the ideal $\text{in}(I)$ is finitely generated.

Definition 5 [**Gröbner basis**] Let $>$ be a multiplicative order on S. A set $G = \{g_1, \ldots, g_s\}$ is called a Gröbner basis for the ideal I (with respect to $>$) if

(a) $g_i \in I$, for each $i = 1, \ldots, s$, and

(b) $\{\text{in}(g_1), \ldots, \text{in}(g_s)\}$ generates the ideal $\text{in}(I)$.

Gröbner Bases are also called *standard bases* by some authors. In particular, this terminology is used in our computer algebra system Macaulay.

Example 6 Let $S = k[a, b, c]$, and let $>$ be the graded reverse lexicographic order, where $a > b > c$. If $I = (a^2 - ac, ab - c^2)$, then

$$G = \{a^2 - ac, ab - c^2, ac^2 - c^3, bc^3 - c^4\}$$

is a Gröbner basis for I. In this case $\text{in}(I) = (a^2, ab, ac^2, bc^3)$. Notice that the highest degree of any element of this Gröbner basis is higher than the highest degree of the original generators, and that there are also more generators for $\text{in}(I)$ than for I.

In fact, there are examples of ideals in 30 variables, generated by 15 quadrics, where the number of Gröbner basis elements is in the thousands.

By modifying the definition of monomial, and of in(f), it is possible to define the Gröbner basis of a submodule $I \subset S^m$, where m is a positive integer. We don't present the details here. (See Bayer's thesis [1]).

Recall that a *syzygy* of the polynomials $\{f_1, \ldots, f_r\}$ is a vector $(g_1, \ldots g_r) \in S^r$ such that $\sum_i f_i g_i = 0$. The syzygy module is the S-module consisting of all such syzygies.

As mentioned above, it is possible to compute the syzygy module of a set of polynomials (or of elements of S^m), (Spear [15], Schreyer [13]). In this lecture, we will not go into the details of how Gröbner bases and syzygies are actually computed. (See [1]) for an exposition of the algorithms). The folowing summarizes what we can compute:

Proposition 7 *(Buchberger, Spear, Schreyer) Let $S = k[x_1, \ldots, x_n]$. Given a submodule $I = (f_1, \ldots, f_r) \subset S^m$, and a multiplicative order $>$ on the monomials of S^m, we can construct*

(a) A Gröbner basis for I, and

(b) A basis for the syzygy module of $\{f_1, \ldots, f_r\}$.

The actual monomial order used makes a significant difference on the performance of the algorithm. In practice, using a reverse lexicographic order generally takes the least amount of time, and the resulting Gröbner bases tend to be the smallest. On the other hand, the lexicographic order tends to take the most time, and the resulting Gröbner bases are often extremely large, and have generators in very high degrees.

3 Basic building blocks

In the rest of the first lecture, we describe three basic building blocks. The first deals with elimination of variables, the second with finding numerical information about ideals or modules, and the third with computing syzygies. At the end, we mention two or three other basic operations, which are also important, but tend to have fewer applications. Currently, almost every application of Gröbner bases to commutative algebra and algebraic geometry can be described in terms of these basic operations.

Before discussing these operations, it is necessary to describe what objects we are interested in.

All of our computations will deal with matrices over certain rings. Recall that an *affine ring* is a ring R which is a finitely generated k-algebra, where k is a field; in other words, $R = k[x_1, x_2, \ldots, x_n]/I$ for some ideal I of $k[x_1, x_2, \ldots, x_n]$.

Most of the rings which we will consider are affine. However, it is possible to extend the computations in two important directions: to the case when k is a P.I.D, and the case where R is a localization of an affine ring. We will not pursue these extensions here.

The objects of greatest interest for us are maps between (affine) rings, modules over such rings, and maps between modules.

3.1 Elimination, and the kernel of a ring map

One of the earliest applications of Gröbner bases was to computing the intersection $I \cap k[x_i, \ldots, x_n]$, for an ideal $I \subset k[x_1, \ldots, x_n]$. Algebraically, this operation eliminates the variables $x_1, \ldots, x_{i-1}$ from I. Geometrically, this contracted ideal defines the image of $Z(I) \subset \mathbf{A}^n$ in $\mathbf{A}^{n-i+1}$, under the projection $(x_1, \ldots, x_n) \mapsto (x_i, \ldots, x_n)$.

In any case, this is easily computed using a special monomial order, and then finding a Gröbner basis:

Proposition 8 *If $>$ is a monomial order on S such that whenever* $\mathrm{in}(f) \in k[x_i, \ldots, x_n]$, *then $f \in k[x_i, \ldots, x_n]$, and if G is a Gröbner basis for I using this order, then $G \cap k[x_i, \ldots, x_n]$ is a Gröbner basis (and hence a generating set) of $I \cap k[x_i, \ldots, x_n]$.*

One example of such an order is the (non-graded) lexicographic order.

Throughout the last decade, several tricks have been used to apply this computation to many situations. However, they almost all fit into the following category.

Let $\phi : A \to B$ be a homomorphism of affine rings. The problem is to find the kernel of this map. This can be reduced to the above problem: If $A = k[x_1, \ldots, x_m]/I$, and $B = k[y_1, \ldots, y_n]/J$, then ϕ is determined by elements $f_1, \ldots f_m \in k[y_1, \ldots, y_n]$, where $\phi(x_i) = f_i \bmod J$.

Proposition 9 *With notation as above,* $\ker \phi$ *is generated by the image of the ideal*

$$(J + (x_1 - f_1(y), x_2 - f_2(y), \ldots, x_m - f_m(y))) \cap k[x_1, \ldots, x_m]$$

in the ring A.

Example 10 **[Subrings, the image of a map to projective space]** Let B be an affine ring. If $f_1, \ldots, f_m \in R$, and $A = k[f_1, \ldots, f_m]$ is the subring generated by these elements, and $\phi : k[x_1, \ldots, x_m] \to B$ is the map sending x_i to f_i, then

$$A \cong k[x_1, \ldots, x_m]/\ker \phi.$$

For example, if $B = k[s,t]$, and $A = k[s^4, s^3t, st^3, t^4]$, then

$$A \cong k[x_1, x_2, x_3, x_4]/(x_2x_3 - x_1x_4,\ x_3^3 - x_2x_4^2,\ x_2^3 - x_1^2x_3,\ x_1x_3^2 - x_2^2x_4).$$

This is the homogeneous coordinate ring defining the image of the map from $\mathbf{P}^1 \to \mathbf{P}^3$ given by $(s,t) \mapsto (s^4, s^3t, st^3, t^4)$.

Example 11 [The secant locus of a variety] This last example can be used to compute many interesting geometric constructions. For example, if $X \subset \mathbf{P}^{n-1}$ is the projective variety corresponding to the graded ring $R = k[x_1, \ldots, x_n]/I$, the secant locus of X, $\mathrm{Sec}(X)$, is by definition the (closure of the) union of all secant lines of X in $\mathbf{P}^{n-1}$.

The secant locus is the closure of the image of the polynomial map:

$$X \times X \times \mathbf{P}^1 \longrightarrow \mathbf{P}^{n-1}$$

given by

$$(\mathbf{p}, \mathbf{q}, (s,t)) \longmapsto (t\mathbf{p} + s\mathbf{q}).$$

Therefore the ideal defining this locus is the kernel of the ring map:

$$k[z_1, \ldots, z_n] \to k[x_1, \ldots, x_n, y_1, \ldots, y_n, t]/K = R \otimes R \otimes k[t],$$

sending z_i to $tx_i + sy_i$, where $K = I(x) + I(y)$.

For example, if $X \subset \mathbf{P}^5$ is the Veronese surface, defined by the ideal, I, in $k[a,b,c,d,e,f]$ generated by the 2×2 minors of the matrix

$$M = \begin{pmatrix} a & b & c \\ b & d & e \\ c & e & f \end{pmatrix},$$

The ideal of the secant locus is obtained by intersecting the ideal

$$I_1 + I_2 + (a - ta_1 - sa_2, \ldots, f - tf_1 - sf_2)$$

in $k[a, \ldots, f, a_1, \ldots, f_1, a_2, \ldots, f_2, s, t]$ with the subring $k[a,b,c,d,e,f]$. (Here, I_1 and I_2 are both the ideal I with each variable $a, \ldots, f$ replaced with $a_i, \ldots, f_i$, for $i = 1, 2$.)

The ideal of the secant locus of the Veronese is generated by the determinant of the matrix M.

Example 12 [The blow up ring, and normal cone] Let R be an affine ring, defining an affine variety X. Let $I \subset R$ be an ideal of R, defining the subvariety Y. (In fact, these may be affine schemes, not just varieties). The *blowup ring* of R along I, or of X along Y, is the ring

$$R \oplus I \oplus I^2 \oplus I^3 \oplus \ldots.$$

This is also the ring $R[It] \subset R[t]$, for a new variable t. A presentation of this ring can be found using subrings, since $R[It]$ is the subring generated by R and $f_1 t, \ldots, f_r t$, if $I = (f_1, \ldots, f_r)$.

The *normal cone* of Y in X is defined to be the ring

$$R[It] \otimes R/I = R/I \oplus I/I^2 \oplus I^2/I^3 \oplus \ldots.$$

If the blowup ring is presented by

$$R[It] = k[x_1, \ldots, x_n, z_1, \ldots, z_r]/K,$$

for some ideal K, then the normal cone ring is simply

$$R[It] \otimes R/I = k[x_1, \ldots, x_n, z_1, \ldots, z_r]/(K + I).$$

Finding the kernel of a ring map $\phi : A \to B$ is a useful basic operation if you are always dealing with ideals and rings. However, it doesn't apply well to modules. In this case, the key operation (of which the kernel is a special case) is the *push forward* of a B-module M:

Definition 13 [pushforward$_1$] Suppose A and B are two affine rings, $\phi : A \to B$ is a ring homomorphism, and b is an $m \times r$ matrix over the ring B. If M is the B-module presented by the matrix b, define $a =$ **pushforward$_1$**(b, ϕ) to be an $m \times s$ matrix over the ring A such that the A-module presented by a is isomorphic to the A-submodule of M generated by the given m generators of M.

For example, if $M = B$ is presented by the 1 by 1 zero matrix $b = (0)$, and $\phi : A \to B$, then **pushforward$_1$**(b, ϕ) is a row vector whose entries generate the kernel of ϕ.

We won't pursue this here, but **pushforward$_1$** can be used to intersect a submodule with a free module of a subring, and to find the conductor of a finite map.

3.2 Numerical invariants, and Hilbert functions

Numerical invariants of rings, ideals and modules are very important. Some numerical information which is important includes the dimension, degree and arithmetic genus of a projective variety, the dimensions of various cohomology groups, the Hodge numbers $h^{p,q}$ of a variety, and Chern classes of vector bundles on projective space.

Hilbert functions form the numerical basis for computing all of these numbers.

Let $S = k[x_1, \ldots, x_n]$ be a graded polynomial ring, where each variable has weight one. Recall that the Hilbert series of a graded S-module M is the function

$$H_M(t) = \sum_{d=-\infty}^{\infty} \dim_k(M_d)t^d,$$

where M_d is the (finite dimensional) k-vector space of all (homogeneous) elements of degree d in M. It is well-known that H_M is a rational function in t, in fact,

$$H_M(t) = P(t)/(1-t)^n,$$

where $P(t)$ is a polynomial, and n is the number of variables of S. Similarly, for large d, there exists a polynomial $h_M(d)$, the Hilbert polynomial, such that for large d, $h_M(d) = \dim_k(M_d)$.

The key relating Hilbert series to Gröbner bases is the following

Proposition 14 *Let $>$ be a multiplicative order on S^m, and suppose that $I \subset S^m$ is a (graded) submodule. Then*

$$\dim(S^m/I)_d = \dim(S^m/\text{in}(I)),$$

and therefore S/I and $S^m/\text{in}(I)$ have identical Hilbert series.

This simple result, originally due to Macaulay, together with a Gröbner basis reduces the problem of finding the Hilbert series of an arbitrary submodule to finding the Hilbert series of a submodule generated by monomials. This becomes a combinatorial problem. For details, see [3] and [12].

It is possible to directly determine the codimension of an ideal (or submodule) from a Gröbner basis without computing the Hilbert series.

The complexity of computing the Hilbert polynomial of a monomial ideal appears to be exponential. In practice, with Sun workstations in 1990, computing the Hilbert polynomial of a monomial ideal with 1000 generators in 30 or 40 variables is often not possible. On the other hand, computing the codimension and degree of such an ideal often takes only a few seconds.

Given the Hilbert series it is easy to write down the codimension, degree, and Hilbert polynomial of a graded module M.

Proposition 15 *If the Hilbert series of a graded S-module M is*

$$H_M(t) = \frac{Q(t)}{(1-t)^{d+1}},$$

where $Q(1) \neq 0$, and $Q(t) = \sum_i a_i t^i$, then

(a) the (affine) dimension of M is $d+1$,

(b) the degree of M is $Q(1)$, and

(c) the Hilbert polynomial of M is

$$h_M(z) = \sum_i a_i \binom{z-i+d}{d}.$$

3.3 Syzygies, and the modulo command

Suppose R is a ring, and $I = (f_1, \ldots, f_r) \subset R^m$ is a submodule of the free module R^m. Recall that the *syzygy module* of I is the submodule of R^r:

$$\{(g_1, \ldots, g_r) \in R^r \mid \sum_{i=1}^{r} f_i g_i = 0\}.$$

The syzygy module of I is also the kernel of the R-homomorphism

$$R^r \xrightarrow{\phi} R^m$$

taking the ith basis vector of R^r to $f_i \in R^m$.

Syzygies are the key to linear algebra over the ring R, and therefore the key to many questions in homological algebra (which is after all simply linear algebra). A slight refinement enables us to easily write down solutions to many computational problems:

Definition 16 **[modulo]** Let a and b be matrices over the ring R, having sizes $r \times m$ and $s \times m$ respectively. We denote by **modulo**(a, b) any matrix over R whose columns generate the kernel of the map

$$R^r \xrightarrow{a} R^m/\mathrm{im}(b),$$

where $\mathrm{im}(b)$ is the submodule generated by the columns of b.

Proposition 17 *Given matrices a and b over an affine ring R as above, there exists an algorithm for finding* **modulo**(a, b).

Proof. If $R = k[x_1, \ldots, x_n]/(h_1, \ldots, h_p)$, and a and b are lifted to matrices A, B, over $S = k[x_1, \ldots, x_n]$, define a matrix of size $m \times (r + s + pm)$ over S by

$$D = (\, A \;\mid\; B \;\mid\; C\,),$$

where if $H = (\, h_1, h_2, \ldots, h_p\,)$, then $C = H \oplus H \oplus \ldots \oplus H$ (m copies of H).

Compute a basis of syzygies on the columns of the matrix D. Each such syzygy $\{\alpha_1, \ldots, \alpha_r, \beta_1, \ldots, \beta_s, \ldots\}$ corresponds to a column vector $\{\alpha_1, \ldots, \alpha_r\}$, and these column vectors generate the kernel of the map $a : R^r \to R^m/\mathrm{im}(b)$, as required. □

In practice, a computer algebra system need not compute the entire syzygy module on the matrix D: it is possible to find **modulo**(a, b) with less work by only keeping track of "partial syzygies."

In the rest of this section, we describe some applications of **modulo**.

Example 18 **[Intersections of ideals]** Let R be an affine ring, and let $I, J, K \subset R$ be ideals. One can compute generators for the intersection $L = I \cap J \cap K$ in the following way. L is the kernel of the R-module homomorphism $\phi : R \to R/I \oplus R/J \oplus R/K$ which sends 1 to $(1, 1, 1)$. If I, J, K are generated by the elements of the row vectors a, b, c, respectively, then L is generated by the elements of the row vector

$$\mathbf{modulo}(\begin{pmatrix} 1 \\ 1 \\ 1 \end{pmatrix}, a \oplus b \oplus c),$$

where

$$a \oplus b \oplus c = \begin{pmatrix} a & 0 & 0 \\ 0 & b & 0 \\ 0 & 0 & c \end{pmatrix}.$$

This construction can be generalized to compute the intersection of any number of submodules of a given module.

Example 19 [Ideal quotients] Let R be an affine ring, and let $I = (a_1, \ldots, a_r)$, $J = (b_1, \ldots b_s)$ be ideals of R. The *ideal quotient*, $(I : J)$ is defined to be the ideal

$$(I : J) := \{g \in R \mid gJ \subset I\}.$$

Ideal quotients are generally useful. In the second lecture, they are used in the construction of the radical of an ideal. They are also used in algebraic geometry for liaison of varieties.

Notice that $(I : J)$ is the kernel of the map

$$R \to R/I \oplus R/I \oplus \ldots \oplus R/I \quad (s\text{copies}),$$

where $1 \mapsto (b_1, \ldots, b_s)$. Therefore, $(I : J)$ is generated by the columns of the matrix

$$\mathbf{modulo}(b^*, a \oplus a \oplus \ldots \oplus a),$$

where b^* is the transpose of b, and there are s copies of a in the second argument of **modulo**.

Example 20 [Annihilators] Let $M = F/\text{im}(a)$ be the cokernel of the matrix $a : G \to F$, where F, G are free R-modules. The *annihilator* of M is the ideal

$$\text{Ann } M := \{g \in R \mid gM = 0\}.$$

For example, if

$$a = \begin{pmatrix} x & y & z \\ y & z & w \end{pmatrix},$$

then $f \in \text{Ann } M$ if and only if $fe_1 = fe_2 = 0$, where $e_1 = \begin{pmatrix} 1 \\ 0 \end{pmatrix}$, $e_2 = \begin{pmatrix} 0 \\ 1 \end{pmatrix}$ are the two generators of M. Consequently, the elements of the annihilator correspond to the first element of each syzygy of the columns of:

$$\begin{pmatrix} 1 & x & y & z & 0 & 0 & 0 \\ 0 & y & z & w & 0 & 0 & 0 \\ 0 & 0 & 0 & 0 & x & y & z \\ 1 & 0 & 0 & 0 & y & z & w \end{pmatrix}.$$

(If f is the first element of such a syzygy, the first two rows tell us that $fe_1 \in \text{im}(a)$, and last two rows tell us that $fe_2 \in \text{im}(a)$).

Another way of viewing this is to use tensor products. This also leads to a more succinct notation. The tensor product of $a : R^3 \to R^2$ and the identity $R^2 \to R^2$, which we denote by $a \otimes R^2$, is the map $R^3 \otimes R^2 \to R^2 \otimes R^2$ given in coordinates by the last 6 columns of the above matrix. The first column of the above matrix is the coordinate matrix of the adjoint of the identity map on R^2, $\Delta : R \to R^2 \otimes (R^2)^*$, where $(R^2)^*$ is the dual free module of R^2. So finally, the annihilator of M in this example is generated by the columns of the matrix $\textbf{modulo}(\Delta, a \otimes (R^2)^*)$. (In this example, the annihilator is generated by the 2×2 minors of the matrix a.)

In the general case, where $a : G \to F$, let

$$\Delta : R \to F \otimes F^*,$$

be the adjoint of the identity on F, and

$$a \otimes F^* : G \otimes F^* \to F \otimes F^*$$

the tensor product of a with the identity on F^*. In this case, the annihilator of $\text{coker}(a)$ is generated by the entries of the row vector

$$\textbf{modulo}(\Delta, a \otimes F^*).$$

This is a relatively simple formula. It also has the advantage of being easy to program by computer, assuming that **modulo** is available. The reason is that the matrices for Δ and $a \otimes F^*$ are straightforward to construct, and involve no computation.

Example 21 [Ext modules] Let M be the cokernel of the matrix $a : F_1 \to F_0$, where F_1, F_0 are free R-modules. Using **modulo** one can easily find a free resolution of M: Define $\textbf{syz}(a) := \textbf{modulo}(a, 0)$ to be a matrix whose columns generate $\ker(a)$. Let $a_1 := a$, and $a_{i+1} := \textbf{syz}(a_i)$, for each $i \geq 1$. The matrices $a_1, a_2, \ldots$ define a free resolution of M:

$$\ldots \longrightarrow F_k \xrightarrow{a_k} F_{k-1} \longrightarrow \ldots \longrightarrow F_1 \xrightarrow{a} F_0,$$

where $a_j : F_j \to F_{j-1}$, and each F_j is a free R-module.

By definition,

$$\text{Ext}^j(M, R) = \ker d^*_{j+1} / \text{im } d^*_j,$$

where d_j^* is the transposed matrix.

This quotient can be written in terms of **modulo**: it is the module which is the cokernel of the matrix

$$\mathbf{modulo}(\mathbf{syz}(d_{j+1}^*), d_j^*).$$

A large number of constructions involving modules or homological algebra can be stated simply using **modulo**, tensor products, and transposes. For example, if M is the cokernel of the matrix $a : F_1 \to F_0$, and N is the cokernel of the matrix $b : G_1 \to G_0$, then $\mathrm{Hom}_R(M, N)$ is presented by the matrix

$$\mathbf{modulo}(\mathbf{modulo}(a^* \otimes G_0, F_1^* \otimes b), F_0^* \otimes b).$$

3.4 Other basic commands

There are two other important techniques, which we don't have time to describe in any detail. The first is the *saturation* of a submodule with respect to an ideal. See either [2] or the forthcoming paper/book with D. Eisenbud [7].

Definition 22 [**sat**] Suppose that a and b are matrices over an affine ring R, where a has size $m \times r$, and b has size $1 \times s$. If $I = \mathrm{im}(a) \subset R^m$ and $J = \mathrm{im}(b) \subset R$, define $\mathbf{sat}(a, b)$ to be a matrix with m rows such that the columns generate the submodule

$$(I : J^\infty) = \{g \in R^m \mid g\, J^N \subset I \text{ for some } N\}.$$

The last technique is extremely useful when analyzing the fibers of a morphism of varieties (ie, a map of rings). For more details, see [2].

Definition 23 [**flatten**] Suppose that R, S are affine rings, that $\phi : R \to S$ is a ring homomorphism, and that M is an S-module presented by a $m \times r$ matrix a. Define $\mathbf{flatten}(a, \phi)$ to be an element $h \in R$ such that the localized module M_h is flat over R_h.

4 Lecture # 2: Finding the radical of an ideal

In this talk we present an algorithm for computing the radical of an ideal, due to Eisenbud and Huneke ([6]).

Let $I \subset S = k[x_1, \ldots, x_n]$ be an ideal in a polynomial ring. Throughout this talk we assume for simplicity that k is an algebraically closed field, although all proofs can be easily extended to the case when k is a perfect field.

The ideal I has an irredundant primary decomposition

$$I = Q_1 \cap Q_2 \cap \ldots \cap Q_s,$$

where Q_i is a P_i-primary ideal, and P_i is a prime ideal of S, for each i. The prime ideals $P_1, P_2, \ldots, P_s$ are the *associated primes* of I.

Geometrically, the zero set of I, $Z(I) = \mathrm{Spec}(S/I) \subset \mathbf{A}^n$ consists of the union of s components, $X_i = \mathrm{Spec}(S/Q_i)$. Some of these X_i are possibly embedded components.

In this talk we present algorithms, due to Eisenbud and Huneke, for computing the following objects.

1. Let $\mathrm{Top}_e(I) := \cap_i Q_i$ be the intersection of all primary components of I having dimension at least e. In the picture above, $\mathrm{Top}_2(I)$ is the defining ideal for the union of the two surfaces. If $d = \dim I$, we write $\mathrm{Top}(I) := \mathrm{Top}_d(I)$. (Here the dimension of an ideal I is by definition $\dim S/I$.)

2. The radical of I is the ideal

   $$\sqrt{I} = \{g \in S \mid g^N \in I, \text{for some } N\}.$$

   This ideal is also the intersection $\cap_i P_i$ of all the associated primes of I.

There are several methods in the literature for computing these ideals, and primary decompositions of ideals (see Seidenberg [14], Gianni-Trager-Zacharias [8], Logar [10], and Bayer-Galligo-Stillman [2]). Thus it is a fair question to ask: why a new method? All of the methods cited above critically depend on finding the equations of the image of a variety (or scheme) under a fairly generic projection. In practice it turns out that there is a large class of problems for which it is essentially impossible to find this image of a reasonably generic projection: often the resulting ideal is generated by a single polynomial having a very large degree. In fact, Eisenbud and Huneke developed this method precisely because they had some examples for which they could not use the above methods.

The general philosophy is that in many cases, it is very easy to compute syzygies (and so finite free resolutions, and ideal quotients), but very difficult to compute projections. The algorithms presented here use syzygy computations but not a single projection.

5 The radical of an ideal

In this section, we will assume that $I \subset S$ is an *equi-dimensional* ideal, having dimension d. Eisenbud and Huneke's method for determining the radical of I is based on the analysis of the singular locus of I.

Let $R = S/I = k[x_1, \ldots, x_n]/(f_1, \ldots, f_r)$. The *Jacobian matrix*, $J(\mathbf{f})$, of $(f_1, \ldots, f_r)$ is the n by r matrix whose (i,j)th entry is $\partial f_j / \partial x_i$.

Define a sequence of ideals

$$I \subset J_d(I) \subset J_{d+1}(I) \subset \ldots \subset J_{n-1}(I) \subset J_n(I) = (1),$$

by setting

$$J_a(I) := I + (\text{all } (n-a) \times (n-a) \text{ minors of the matrix } J(\mathbf{f}))$$

for each $d \leq a \leq n-1$. By convention, we set $J_n(I) = (1)$, and $\dim J_n(I) = -1$.

A point $p \in \mathbf{A}^n$ is a *singular point* of $X = Z(I)$ if $f(p) = 0$ for every $f \in J_d(I)$. The set of all singular points of X is denoted by $\mathrm{Sing}(X)$, or $\mathrm{Sing}(I)$, or even by $\mathrm{Sing}(R)$, if $R = S/I$.

The singular locus is related to the components of I via the following fact.

Proposition 24 *If $I = Q_1 \cap \ldots \cap Q_s$ is an irredundant primary decomposition of the equidimensional ideal I, $X = \mathrm{Spec}\ S/I$, and $X_i = \mathrm{Spec}\ S/Q_i$, for $i = 1, \ldots, s$, then*

$$\mathrm{Sing}(X) = \mathrm{Sing}(X_1) \cup \ldots \cup \mathrm{Sing}(X_s) \cup \bigcup_{i<j} X_i \cap X_j.$$

Proof. Note that $\text{Sing}(X)$ can be identified with the set of maximal ideals, m, of $R = S/I$ such that the localization R_m is not a regular local ring. Given a maximal ideal m containing I, one of the following two cases must hold.

(1) There is a unique index i such that $Q_i \subset m$. In this case $R_m \cong (S/Q_i)_m$, and so $m \in \text{Sing}(X) \iff m \in \text{Sing}(X_i)$.

(2) There are at least two distinct indices i, j such that $Q_i \subset m$ and $Q_j \subset m$. In this case R_m is not a domain, hence not regular, and $m \in X_i \cap X_j$. □

The following example shows the intuition which relates singular loci to radicals.

Example 25 Let $I = (x^2, xy, y^2) \subset k[x, y, z]$ be a non-reduced ideal supported on the z-axis in $\mathbf{A}^3$. In this case, $J_1(I) = I$. In other words, every point of the line is singular. More generally, if Q is P-primary, but $Q \neq P$, then every point of $Z(Q)$ is singular.

Using the above geometric description of the singular locus, together with this example, the following proposition is geometrically clear.

Proposition 26 *If $I \subset S$ is an equi-dimensional ideal of dimension d, then*

$$I = \sqrt{I} \iff \dim J_d(I) < d.$$

Proof. The key ingredient is Serre's criterion for a ring $R = S/I$ to be reduced (i.e. I is a radical ideal): R is reduced if and only if (a) R_P is a regular local ring for every associated prime P of R, and (b) I has no embedded components.

We also apply Serre's theorem: If R is regular, and $P \subset R$ is a prime ideal, then R_P is regular.

Suppose that $I = \sqrt{I}$. Using the same notation as in the previous proposition, this implies $\dim \text{Sing}(X_i) < \dim X_i = d$ (See Hartshorne's book, pg. 33). Since $\dim X_i \cap X_j < d$ as well, it follows that

$$\dim J_d(I) = \dim \text{Sing}(X) < d.$$

Conversely, suppose that $\dim \text{Sing}(X) < d$. For each associated prime P_i of I, there is a maximal ideal $m \supset P_i$ such that $R_m \cong (S/Q_i)_m$ is a regular local ring. By Serre's theorem, this implies that R_{P_i} is a regular local ring. By Serre's criterion, since I has no embedded components, R is reduced. □

This result allows us to easily test whether a given (equidimensional) ideal is radical. On the other hand, if I is not radical, it doesn't tell us how to improve I, i.e. what polynomials to add to I to obtain the radical.

The following key observation of Eisenbud and Huneke solves this problem.

Proposition 27 *Let $I \subset S$ be an equi-dimensional ideal of dimension d. If* $\dim J_a(I) = d$ *and* $\dim J_{a+1}(I) < d$*, for some $a \geq d$, then*

$$\sqrt{I} = \sqrt{I : J_a(I)}.$$

Several remarks: (1) $I \subset (I : J_a(I)) \subset \sqrt{I}$, and $I \neq (I : J_a(I))$, so $(I : J_a(I))$ is an actual improvement of I: it is a larger ideal which is also equidimensional, and has the same radical.

(2) If $\dim J_a(I) < d$, then $(I : J_a(I)) = I$.

Example 28 If $\dim J_a(I) = d$, it is not necessarily the case that I and $(I : J_a(I))$ have the same radical: it is important also that $\dim J_{a+1}(I) < d$. For example, let $I = (x, y, z)^3 \subset k[x, y, z]$. One computes easily that

$$\begin{aligned} J_0(I) &= (x, y, z)^3 \\ J_1(I) &= (x, y, z)^3 \\ J_2(I) &= (x, y, z)^2 \\ J_3(I) &= (1). \end{aligned}$$

Notice that $(I : J_0(I)) = (I : J_1(I)) = (1)$, but $(I : J_2(I)) = (x, y, z)$.

Using the above two propositions, we have a particularly simple algorithm for actually computing the radical of an equidimensional ideal.

Algorithm 29 (The radical of an equidimensional ideal)

input:
  an ideal $I \subset k[x_1, \ldots, x_n]$, equidimensional of dimension d.
output:
  The radical, $\sqrt{I}$, of I.
begin
  compute $J_d(I)$.
  if $\dim J_d(I) < d$
    then return I.
  Find the unique integer a, $d \leq a \leq n-1$ such that
    $\dim J_a(I) = d$, and $\dim J_{a+1}(I) < d$.
  return radical$(I : J_a(I))$.
end.

In order to use this algorithm, we must be able to compute the determinants of the minors of a Jacobian matrix, the dimension of an ideal, and ideal quotients. Conceptually, computing the determinants is simple, and we have already seen how to compute the dimension of an ideal using Gröbner bases, and how to compute ideal quotients using syzygies. David Eisenbud has implemented this algorithm in the Macaulay system.

Notice that if $\dim J_{a+1}(I) < d$, and $I_1 := (I : J_a(I))$, then $J_{a+1}(I) \subset J_{a+1}(I_1)$. This means that the value "a" for the ideal I_1 is no larger than the value "a" for the ideal I.

Example 30 Let $I = (f^{100}) \subset k[x_1, \ldots, x_n]$, where f is a squarefree polynomial. We compute by hand that

$$J_{n-1}(I) = (f^{99}\partial f/\partial x_1, \ldots, f^{99}\partial f/\partial x_n, f^{100})$$

and that $(I : J_{n-1}(I)) = (f)$. This method finds the radical in one step. In a similar way, if $I = (f)$, for any polynomial f, this algorithm computes the squarefree part of f (i.e. the radical of I) in one step.

Example 31 For this example, see the Macaulay listing on page E-1. Let $I \subset k[a, b, c, d, e]$ be the ideal generated by the entries of the matrix

$$\begin{pmatrix} a+b+c+d+e \\ ab+bc+cd+de+ea \\ abc+bcd+cde+dea+eab \\ abcd+bcde+cdea+deab+eabc \end{pmatrix}$$

Using Macaulay, one discovers that codim $I = 4$, and so the dimension of I is 1. Since there are four generators for I, this ideal is a complete intersection, and hence it is equidimensional. Applying the algorithm, we find that $\dim J_1(I) = 1$, and $\dim J_2(I) = 0$, so that $a = 1$.

After computing $I_1 := (I : J_1(I))$ (see the Macaulay listing), we find that $\dim J_2(I_1) = 0$, and so I_1 is the radical of I.

This entire computation takes about 30 seconds on a Sun Sparcstation.

Example 32 For this example, refer to the Macaulay session on pages E-2 and E-3. Let $I \subset k[a,b,c,d]$ be the ideal generated by 2 sextics and 3 polynomials of degree 9 on that page. I is the link of the rational quartic curve in $\mathbf{P}^3$ by a non-reduced complete intersection of degree 36. Therefore I is an equidimensional ideal of (affine) dimension 2, and degree 32.

Using Macaulay, we find that $\dim J_3(I) = 2$. Since $\dim J_4(I) = -1$, we compute $I_1 := (I : J_3(I))$. This ideal is displayed at the bottom of page E-2.

Next, we compute that $\dim J_3(I_1) = 1$, but $\dim J_2(I_1) = 2$. Therefore we let $I_2 := (I_1 : J_2(I_1))$. Finally, we compute that $\dim J_2(I_2) = 1 < 2$, and so I_2 is the radical of I. This ideal has degree 14, and is reproduced on page E-3.

We now turn to the proof of Proposition 27.

The cokernel of the Jacobian matrix, when reduced by I, is the module of Kähler differentials Ω_R:

$$R^r \stackrel{J(\mathbf{f}) \bmod I}{\longrightarrow} R^n \longrightarrow \Omega_R \longrightarrow 0.$$

For each a, define $J_a(R)$ to be the image of the ideal $J_a(I)$ in R.

The key fact relating Ω_R to the singularities of R is

Fact 33 *Let R be a finitely generated k-algebra (where k is a perfect field). Then*

$$\begin{aligned} & R \text{ is smooth of dimension } d \\ \iff & \Omega_R \text{ is a free } R\text{-module of rank } d \\ \iff & J_d(I) = (1), \text{and } J_{d-1}(I) = I. \end{aligned}$$

This last equivalence follows from facts about Fitting ideals of modules. For more details, see Matsumura's book [11], and the paper [5] of Buchsbaum and Eisenbud.

Proof. (of Proposition 27) To prove the proposition, let P be an associated prime of I. We need to show that it is also an associated prime of $I_1 := (I : J_a(I))$. Suppose that this is not the case. Then $(I : J_a(I)) \not\subset P$, and therefore $J_a(R)_P = 0$. By hypothesis, $J_{a+1}(R) \not\subset P$, and so $J_{a+1}(R)_P = R_P$. The lemma below then implies that there is a $g \in R$ such that $d = \dim R = \dim R_g = a + 1 > d$, a contradiction. □

Lemma 34 *R is generically smooth along P, of dimension $a+1$ if and only if $J_a(R)_P = 0$ and $J_{a+1}(R)_P = (1)$.*

Proof. By definition, R is generically smooth along P, of dimension $a + 1$ if and only if there exists a $g \notin P$, such that the affine localization R_g is smooth of dimension $a + 1$.

Likewise, there exists such a g if and only if $J_{a+1}(R_g) = (1)$, and $J_a(R_g) = 0$

□

6 Ext and removing low dimensional components

Let $I \subset S$ be an ideal. In this section we describe Eisenbud and Huneke's method for finding $\mathrm{Top}_e(I)$. The general philosophy is the same as for computing radicals: Use as many syzygy computations as needed, but do not use projections.

The observation that Eisenbud and Huneke make is that the associated primes of I can all be detected using the S-modules $\mathrm{Ext}^j(S/I, S)$. In fact, the associated primes all "percolate" to the top, in a sense. For the following proposition, recall that the codimension of a S-module M is by definition the codimension of its annihilator ideal, $\mathrm{Ann}(M)$, in S. Similarly, the associated primes of M are those of the ideal $\mathrm{Ann}(M)$.

Proposition 35 *Let $I \subset S$ be an ideal. Then*

(a) If $\mathrm{Ext}^j(S/I, S) \neq 0$, then $\mathrm{codim}\ \mathrm{Ext}^j(S/I, S) \geq j$, and

(b) The set of associated primes of $\mathrm{Ext}^j(S/I, S)$ of codimension j is precisely the set of associated primes of I of codimension j.

Proof. There are two main ingredients to the proof. The first is Serre's theorem: If R is a regular local ring, and $P \subset R$ is a prime ideal, the localization R_P is also regular.

The second is the formula of Auslander and Buchsbaum: If R is a regular local ring, then for every finitely generated R-module M,

$$\mathrm{pd}_R(M) + \text{depth } M = \dim R.$$

The only fact we need about depth is that depth $R_P = 0$ if and only if P is an associated prime of R.

To prove (a), note that if codim $P < j$, then $\mathrm{Ext}^j(S/I, S)_P = 0$, since $\mathrm{pd}(S/I)_P \leq \text{codim } P < j$.

For (b), let P be a prime ideal of S having codimension j. Then

$$\begin{aligned} & P \text{ is an associated prime of } \mathrm{Ext}^j(S/I,S) \\ \iff & \mathrm{Ext}^j(S/I,S)_P \neq 0 \\ \iff & \mathrm{pd}_{S_P}(S/I)_P = j \\ \iff & \mathrm{depth}(S/I)_P = 0 \\ \iff & P \text{ is an associated prime of } I. \end{aligned}$$

□

Algorithm 36 ($\mathrm{Top}_e(I)$)

```
input:
    an ideal I ⊂ S = k[x_1, ..., x_n], and an integer e ≥ 0.
output:
    Top_e(I).
begin
    compute a free resolution of I.
    for each j from n down to n − e + 1 do (*)
        compute Ext^j(S/I, S) (**).
        if codim Ext^j(S/I, S) = j then
            compute J := Ann(Ext^j(S/I, S)).
            set I := (I : J).
    return I.
end.
```

In order to use this algorithm, we must be able to compute $\mathrm{Ext}^j(S/I, S)$, the annihilator of a module, ideal quotients, and the codimension of an ideal.

In the first talk, we have shown how to compute these using Gröbner bases and syzygies.

In this algorithm, the loop (*) can be done in any order. In doesn't matter what codimension j is done first. The only important point is that all of codimensions $j = n - e + 1, \ldots, n$ be done. In practice using Macaulay, it seems that starting with $j = n$ and proceeding downward is more efficient.

Also, in statement (**), one can either use the original I or the new I to compute the Ext^j module, without affecting the correctness of the algorithm. In practice, it appears to be faster in general to recompute the Ext modules; the resulting simplification of the ideal often repays the extra cost of recomputing the necessary finite free resolutions.

Example 37 For this example, refer to the Macaulay listing on page E-4.

References

[1] D. Bayer, *The division algorithm and the Hilbert scheme*, Ph.D. thesis, Harvard University (1982). Order number 82-22588, University Microfilms International, 300 N. Zeeb Rd., Ann Arbor, MI 48106.

[2] D. Bayer, A. Galligo, M. Stillman, *Primary Decompositions of ideals*, in preparation.

[3] D. Bayer and M. Stillman, *The computation of Hilbert functions*, in preparation.

[4] B. Buchberger, *Ein algorithmus zum Auffinden der Basiselemente des Restklassenringes nach einem nulldimensionalen Polynomideal.*, Ph.D. Thesis, Universität Innsbrück (1965).

[5] D. Buchsbaum and D. Eisenbud, *What annihilates a module?* (Find exact reference)

[6] D. Eisenbud and C. Huneke, *A Jacobian method for finding the radical of an ideal*, preprint (1989).

[7] D. Eisenbud and M. Stillman, *Methods for computing in algebraic geometry and commutative algebra*, in preparation.

[8] P. Gianni, B. Trager, and G. Zacharias, *Gröbner bases and primary decompositions of polynomial ideals*, in "Computational Aspects of Commutative Algebra", ed. L. Robbiano, 15-33 (1989).

[9] R. Hartshorne, Algebraic Geometry, Graduate texts in mathematics, no. 52, Springer-Verlag, New York (1977).

[10] A. Logar, *Complexity of computing radicals of ideals*, in preparation.

[11] H. Matsumura, Commutative Algebra, Cambridge University Press, Cambridge (1986).

[12] F. Mora and H. Möller, *Computation of Hilbert functions*, (Find reference).

[13] F. Schreyer, *Die Berechnung von Syzygien mit dem verallgemeinerten Weierstrasschen Divisionsatz*, Diplomarbeit Hamburg, (1980).

[14] A. Seidenberg, *Constructions in algebra*, Trans. Amer. Math. Soc. **197**, 273-313 (1974).

[15] D. Spear, *A constructive approach to commutative ring theory*, Proc. 1977 MACSYMA User's Conference, 369-376.

cyclic5.log Sun Mar 11 23:25:32 1990 1

```
% <Cyclic5

;;; In this example, we find the radical of a complete intersection
;;; whose zero set in P4 consists of points.
;;; We can use the equidimensional radical algorithm since the ideal
;;; is a complete intersection.

1% <ring 5 a-e r

1% <ideal j a+b+c+d+e ab+bc+cd+de+ea abc+bcd+cde+dea+eab abcd+bcde+cdea+deab+eabc

; <jideal is a Macaulay script which finds the ideal Jd(I) as in the talk.
; we first locate a such that J_a(j) has dim = 1, but J_{a+1}(j) has dim = 0.

1% <Jideal j 4 d4
; codimension : 5

1% <Jideal j 3 d3
; codimension : 5

1% <Jideal j 2 d2
; codimension : 5

1% <Jideal j 1 d1
; codimension : 4

; at this point we finally know that the "a" above is 1.

1% quotient j d1 j1
; -5.-4.-3.-2.-1.0.1..2..3..4..5..6..7..
; computation complete after degree 7

; at this point j1 = radical(j), although we don't know that yet.
; The next step verifies this fact.

1% <Jideal j1 1 d1
[252k]; codimension : 5

; since this has codimension 5, j1 is reduced, and we are done.

1% putmat j1
; 1
; 5
; a+b+c+d+e
; b2+bd-cd+2be+ce+e2
; bc2-bcd+c2d-c2e+bde+cde+d2e-be2-2ce2+de2-e3
; bcd2-104/35c2d2-104/35cd3+104/35c3e-67/122bcde+918c2de-111/7bd2e-2750cd2e \
;     -3669d3e+69/35bce2+104/7c2e2+67/122bde2-67/109d2e2+3669be3-42/109ce3 \
;     +67/122de3+3669e4
; c2d2+cd3-c3e-2bcde-c2de+5bd2e+3cd2e+4d3e-bce2-5c2e2+2bde2+8d2e2-4be3-8ce3 \
;     +2de3-4e4

;;; Finding the radical has been implemented by D. Eisenbud in Macaulay:

1% <unmixed_radical j jrad
[315k]
1% putmat jrad
; 1
; 5
; a+b+c+d+e
; b2+bd-cd+2be+ce+e2
; bc2-bcd+c2d-c2e+bde+cde+d2e-be2-2ce2+de2-e3
; bcd2+3/16c2d2+3/16cd3-3/16c3e+5/8bcde+13/16c2de-1/16bd2e+25/16cd2e-1/4d3e \
;     -19/16bce2-15/16c2e2-5/8bde2-1/2d2e2+1/4be3-1/2ce3-5/8de3+1/4e4
; c2d2+cd3-c3e-2bcde-c2de+5bd2e+3cd2e+4d3e-bce2-5c2e2+2bde2+8d2e2-4be3-8ce3 \
;     +2de3-4e4

% exit
```

Find the radical of this ideal

the radical

example2.log Mon Mar 12 13:39:09 1990 1

```
% <example2

;; Example: A non-reduced structure on a link of the rational quartic
;; In this example, we find the radical of a projective curve of degree 32.

1% <ring 4 a-d r

1% <monomial_curve 1 3 4 j
; warning: module isn't homogeneous

1% type j
; c3-bd2 bc-ad b3-a2c ac2-b2d

1% <ideal k (c3-bd2)^2+(bc-ad)^3 (b3-a2c)^2

1% quotient k j I
; 0.1.2.3.4.5.6..7..8..9..10..11..
; computation complete after degree 11

; lets look at our ideal I, the link of the rational quartic,
; whose radical we wish to find.
1% degree I
; codimension : 2
; degree      : 32

1% putmat I
; 1
; 5
; b3c3+c6-3ab2c2d+3a2bcd2-2bc3d2-a3d3+b2d4
; b6-2a2b3c+a4c2
; a2b5c2-a4b2c3-a2b2c5-b2c7-2a3b4cd+2a5bc2d-a3bc4d-4abc6d+a4b3d2-a6cd2 \
;     +8a2b3c2d2+a4c3d2-2c7d2-ab5d3-10a3b2cd3+14ab2c3d3+4a4bd4-b4cd4-6a2bc2d4 \
;     +4bc4d4-4ab3d5+2a3cd5-2b2cd6
; a4bc4+2a2bc6+bc8-a3b3c2d-2a5c3d+a3c5d+4ac7d-a2b5d2+4a4b2cd2+b5c2d2 \
;     -12a2b2c3d2-2b2c5d2-a5bd3+ab4cd3+10a3bc2d3-8abc4d3+a2b3d4-4a4cd4+b3c2d4 \
;     +4ab2cd5
; a5c4+2a3c6+ac8+2a2b5cd-5a4b2c2d-2a2b2c4d+b2c6d-a3b4d2+4a5bcd2-2ab4c2d2 \
;     -2abc5d2-a6d3+4a2b3cd3-2a4c2d3+2c6d3-5ab2c2d4+b4d5+6a2bcd5-4bc3d5-2a3d6 \
;     +2b2d7
```

We want the radical of this ideal.

```
1% <Jideal I 3 d3
; codimension : 2

; since dim J_3(I) = 4-2, and dim J_4(I) < 2
; we now compute (I : J_3(I))

1% quotient I d3 I1
; -2.-1.0.1.2.3.4.5.[252k]6..7..8.[315k].9..
; computation complete after degree 9

; I1 is somewhat simpler, having degree 24:
1% degree I1
; codimension : 2
; degree      : 24

1% putmat I1
; 1
; 5
; b6-2a2b3c+a4c2
; b3c3+c6-3ab2c2d+3a2bcd2-2bc3d2-a3d3+b2d4
; a2b3c2-1/3a4c3+1/3a2c5-1/3c7-1/3ab5d-a3b2cd+5/3ab2c3d+2/3a4bd2-1/3b4cd2 \
;     -4/3a2bc2d2+2/3bc4d2-2/3ab3d3+1/3a3cd3-1/3b2cd4
; ab5c+ab2c4-a2b4d-a4bcd+b4c2d+a5d2-a3c2d2-a2b2d3
; a2b4c-a4bc2-a2bc4-bc6-a3b3d+a5cd+5ab3c2d-2a3c3d+b5d2-4a2b2cd2+2b2c3d2+a3bd3 \
;     -b3d4
```

$(I : J_3(I))$

```
; now find J_3(I1):
1% <Jideal I1 3 d3
; codimension : 3

; and J_2(I1):
1% <Jideal I1 2 d2
; codimension : 2

1% quotient I1 d2 I2
; -6.-5.[441k]-4.-3.[504k]-2.-1.0.1.2.[567k]3..4.[630k].5.[692k].6..7.[755k].8.
; .
```

example2.log   Mon Mar 12 13:39:09 1990   2

```
; computation complete after degree 8

1% degree I2
; codimension : 2
; degree      : 14

1% putmat I2
; 1
; 5
; b3-a2c
; a3c3+ac5-2a2b2cd+b2c3d+a3bd2-abc2d2-a2cd3
; a2bc3+bc5-2a3c2d+ac4d+a2b2d2-b2c2d2-abcd3
; a2c4+c6-3ab2c2d+3a2bcd2-2bc3d2-a3d3+b2d4
; a2b2c2+b2c4-2a3bcd+abc3d+a4d2-a2c2d2-ab2d3
```

$I_2 := (I_1 : J_2(I))$

and $I_2 = \sqrt{I}$

```
; now let's do it again:
1% <Jideal I2 2 d2
; codimension : 3

; since dim J_2(I2) = 1 < 2, I2 is a radical ideal, the radical
; of the original ideal I.

;-------------------------------------------------------
1% <ideal k1 (c3-bd2)^2+(bc-ad)^3 (b3-a2c)

1% quotient k1 j I'
; 0.1.2.3..4..5..6..7..
; computation complete after degree 7

1% degree I'
; codimension : 2
; degree      : 14

1% putmat I'
; 1
; 5
; b3-a2c
; a2b2c2+b2c4-2a3bcd+abc3d+a4d2-a2c2d2-ab2d3
; a2bc3+bc5-2a3c2d+ac4d+a2b2d2-b2c2d2-abcd3
; a3c3+ac5-2a2b2cd+b2c3d+a3bd2-abc2d2-a2cd3
; a2c4+c6-3ab2c2d+3a2bcd2-2bc3d2-a3d3+b2d4

; this is the radical also!
```

(Received: October 1990)

Acta Applicandae Mathematicae **21**: 105–135, 1990.

Combinatorial Algorithms for the Expansion of Various Products of Schur Functions

by

Jeffrey B. Remmel*

Department of Mathematics, University of California, San Diego, U.S.A.

AMS subject classifications (1980). 20B30, 20G05, 05Axx.
Key words. Schur functions, combinatorics of Young tableaux.

*Partially supported by NSF Grant #DMS 87-02473

Introduction

The main purpose of these lectures is first to briefly survey the fundamental connection between the representation theory of the symmetric group $\mathcal{S}_n$ and the theory of symmetric functions and second to show how combinatorial methods that arise naturally in the theory of symmetric functions lead to efficient algorithms to express various products of representations of $\mathcal{S}_n$ in terms of sums of irreducible representations. That is, there is a basic isometry which maps the center of the group algebra of $\mathcal{S}_n$, $Z(\mathcal{S}_n)$, to the space of homogeneous symmetric functions of degree n, Λ^n. This basic isometry is known as the Frobenius map, F. The Frobenius map allows us to reduce calculations involving characters of the symmetric group to calculations involving Schur functions. Now there is a very rich and beautiful theory of the combinatorics of symmetric functions that has been developed in recent years. The combinatorics of symmetric functions, then leads to a number of very efficient algorithms for expanding various products of Schur functions into a sum of Schur functions. Such expansions of products of Schur functions correspond via the Frobenius map to decomposing various products of irreducible representations of $\mathcal{S}_n$ into their irreducible components. In addition, the Schur functions are also the characters of the irreducible polynomial representations of the general linear group over the complex numbers $GL_n(\mathbb{C})$. Thus the combinatorial algorithms for the expansions of Schur functions also have applications for decomposing various representations of $GL_n(\mathbb{C})$ into their irreducible components. In this paper, we will describe some recent work on developing efficient combinatorial algorithms for computing various products of symmetric functions including the ordinary product of Schur functions, the Kronecker product of Schur functions, and the plethysm of Schur functions. The outline of this paper is as follows. In section one, we shall describe the spaces $Z(\mathcal{S}_n)$ and Λ^n and define the fundamental bases for these spaces. In section two, we shall describe the basic products of Schur functions which we shall consider and their connections with the representation theory of $\mathcal{S}_n$ and $GL_n(\mathbb{C})$. Then in sections 3, 4, and 5 we shall outline combinatorial methods for expansion of these products of Schur functions as a sum of Schur functions.

§1. The spaces $Z(\mathcal{S}_n)$ and Λ^n

Let $A(\mathcal{S}_n) = \{f : \mathcal{S}_n \to \mathbb{C}\}$ denote the group algebra of the symmetric group $\mathcal{S}_n$ over the complex numbers $\mathbb{C}$. We shall often write $f \in A(\mathcal{S}_n)$ as a formal sum $f = \sum_{\alpha \in \mathcal{S}_n} f(\sigma)\sigma$. This given, the sum, product, and scalar product on $A(\mathcal{S}_n)$ are defined by

$$f + g = \sum_{\sigma \in \mathcal{S}_n} (f(\sigma) + g(\sigma))\sigma\,,$$

$$f * g = \sum_{\sigma \in \mathcal{S}_n} \left(\sum_{\tau \in \mathcal{S}_n} f(\tau) g(\tau^{-1}\sigma) \right) \sigma\,,$$

$$\langle f, g \rangle = \frac{1}{n!} \sum_{\sigma \in \mathcal{S}_n} f(\sigma)\overline{g(\sigma)}.$$

where $\overline{g(\sigma)}$ denotes the complex conjugate of $g(\sigma)$. It is easy to show that $f \in A(\mathcal{S}_n)$ is in the center of the group algebra of $\mathcal{S}_n$, denoted by $Z(\mathcal{S}_n)$, if and only if f is constant on the conjugacy classes. Now given $\alpha \in \mathcal{S}_n$, we let $\lambda(\alpha)$ denote the (numerical) partition of n which results from the increasing rearrangement of the cycle lengths of α, e.g., if $\alpha = (1456)(28)(379)(10\ 11)$, $\lambda(\alpha) = (2, 2, 3, 4)$. It is well known that α is conjugate to τ in $\mathcal{S}_n$ if and only if $\lambda(\alpha) = \lambda(\tau)$ so that the conjugacy classes of $\mathcal{S}_n$ are naturally indexed by partitions of n. We write $\lambda \vdash n$ if λ is a partition of n. Given a partition $\lambda = (0 < \lambda_1 \le \cdots \le \lambda_k)$, we define $\lambda! = \lambda_1! \cdots \lambda_k!$. We shall also write $\lambda = (1^{\alpha_1} 2^{\alpha_2} \cdots n^{\alpha_n})$ to indicate that λ has α_i parts of size i for $i = 1, 2, \ldots, n$. If $\lambda = (1^{\alpha_1} 2^{\alpha_2} \cdots n^{\alpha_n})$, we let $z_\lambda = \alpha_1! \alpha_2! \cdots \alpha_n! 1^{\alpha_1} 2^{\alpha_2} \cdots n^{\alpha_n}$. The conjugacy class corresponding to $\lambda \vdash n$ is denoted by C_λ. Thus $C_\lambda = \{\sigma \in \mathcal{S}_n | \lambda(\sigma) = \lambda\}$. Then it is well known that $|C_\lambda| = \frac{n!}{z_\lambda}$. We denote by C^λ, the element of the group algebra corresponding to C_λ. That is,

$$C^\lambda = \sum_{\sigma \in \mathcal{S}_n} \chi(\lambda(\sigma) = \lambda)\sigma \tag{1.1}$$

where for any statement S, the indicator $\chi(S) = 1$ if S is true and $\chi(S) = 0$ if S is false. Thus $\{C^\lambda | \lambda \vdash n\}$ forms a natural basis for $Z(\mathcal{S}_n)$.

Next we shall give Young's combinatorial definition of the following three characters

$$p^\lambda = \chi^{ind(1 \uparrow_{\mathcal{S}_{\lambda_1} \times \cdots \times \mathcal{S}_{\lambda_k}}^{\mathcal{S}_n})} \tag{1.2}$$

$$n^\lambda = \chi^{ind(sgn \uparrow_{\mathcal{S}_{\lambda_1} \times \cdots \times \mathcal{S}_{\lambda_k}}^{\mathcal{S}_n})} \tag{1.3}$$

(1.4) $\quad \chi^\lambda =$ irreducible character of $\mathcal{S}_n$ corresponding to the conjugacy class λ.

That is, if $\lambda(0 < \lambda_1 \leq \cdots \leq \lambda_k)$ is a partition of $\mathcal{S}_n$, then $\mathcal{S}_{\lambda_1} \times \cdots \times \mathcal{S}_{\lambda_k}$ denotes the so-called Young subgroup of $\mathcal{S}_n$ consisting of all permutations which permute the elements $1, \ldots, \lambda_1$ among themselves, permute $\lambda_1 + 1, \ldots, \lambda_1 + \lambda_2$ among themselves, permute $\lambda_1 + \lambda_2 + 1, \ldots, \lambda_1 + \lambda_2 + \lambda_3$ among themselves, etc. For any subgroup H of $\mathcal{S}_n$ and representation A of H, we let χ^A denote the character of A and $ind\, A \uparrow {}^{\mathcal{S}_n}_{H}$ denote the induced representation. We also let 1 denote the trivial representation of $\mathcal{S}_n$ and sgn denote the sign representation of $\mathcal{S}_n$. Thus $\chi^1(\sigma) = 1$ for all σ and $\chi^{sgn}(\sigma) = sgn(\sigma)$ for all $\sigma \in \mathcal{S}_n$.

Let $\lambda = (0 < \lambda_1 \leq \cdots \leq \lambda_k)$ be a partition of n. We let λ' denote the conjugate of λ, i.e., λ' is the partition which results by reading the column lengths of F_λ from right to left. A column strict tableau T of shape λ is a filling of F_λ with positive integers in such a way that the numbers weakly increase in each row from left to right and strictly increase in each column from bottom to top. T is called a standard tableau if each $i \in \{1, 2, \ldots, n\}$ occurs once and only once in T. For example, T_1 and T_2 below are examples of a column strict tableau and a standard tableau of shape $(1, 2, 2, 3)$, respectively.

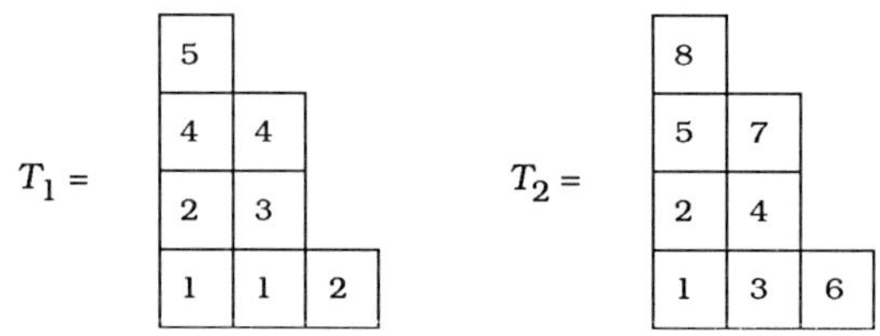

Figure 1.1

Given any subset $\{i_1, \ldots, i_k\}$ of $\{1, \ldots, n\}$, we let

$$[i_1, \ldots, i_k] = \sum_{\sigma \in \mathcal{S}_n} \sigma \chi(\sigma(j) = j \quad \text{if} \quad j \notin \{i_1, \ldots, i_k\})$$

and

$$[i_1, \ldots, i_k]' = \sum_{\sigma \in \mathcal{S}_n} sign(\sigma) \sigma \chi(\sigma(j) = j \quad \text{if} \quad j \notin \{i_1, \ldots, i_k\}).$$

Thus $[i_1, \dots i_k]$ is the element of the group algebra which is 1 on those permutations which permute $i_1, \dots i_k$ among themselves, and leave all other elements fixed. The value of $[i_1, \dots, i_k]$ is 0 on all permutations which do not fix the complement of $\{i_1, \dots, i_k\}$ elementwise. $[i_1, \dots, i_k]'$ is defined similarly except that we take the sign of σ instead of 1. Given a standard tableau T of shape λ, we let $[R_i] = [r_1^i, \dots, r_{\lambda_i}^i]$ where $r_1^i, \dots, r_{\lambda_i}^i$ are the elements in the ith row of T where we label the rows from bottom to top. $[C_i]' = [c_1^i, \dots, c_{t_i}^i]'$ where $c_1^i, \dots, c_{t_i}^i$ are the elements in the ith column of T where we label the columns from left to right. Then we define two elements $P(T)$ and $N(T)$ in the group algebra $A(\mathcal{S}_n)$ by

$$P(T) = \frac{1}{\lambda!}\,[R_1][R_2]\cdots[R_{k(\lambda)}]\,, \tag{1.2}$$

$$N(T) = \frac{1}{\lambda'!}\,[C_1]'[C_2]'\cdots[C_{k(\lambda')}]'\,. \tag{1.3}$$

For example, for T_2 pictured above

$$P(T_2) = \frac{1}{2!2!3!}\,[8][5,7][2,4][1,3,6]\,,$$
$$N(T_2) = \frac{1}{1!3!4!}\,[1,2,5,8]'[3,4,7]'[6]'\,.$$

Finally, we define the *hook length* of cell (i,j) in F_λ, denoted by h_{ij}, to be 1 plus the number of cells to the right or above cell (i,j) in F_λ. Put $\mathbf{h}_\lambda = \prod_{(i,j)\in F_\lambda} h_{ij}$. For example, the hook lengths of the cells of $F_{(1,2,2,3)}$ are given in Figure 1.2.

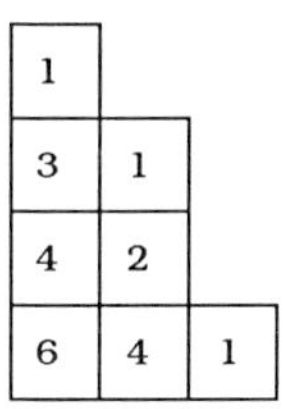

Figure 1.2

Thus $\mathbf{h}_{(1,2,2,3)} = 6 \cdot 4 \cdot 4 \cdot 3 \cdot 2 = 576$. This given, Young defines one more element of the group algebra based on a standard tableau T of shape λ, namely

$$Y(T) = \frac{\lambda!\lambda'!}{\mathbf{h}_\lambda}\,P(T)N(T)\,. \tag{1.4}$$

If we then think of $\sigma \in \mathcal{S}_n$ as the element of the group algebra which is 1 at σ and 0 otherwise, we can define the three Young idempotents mentioned above as follows. Take any standard tableau T of shape λ. Then

$$p^\lambda = \sum_{\sigma \in \mathcal{S}_n} \sigma P(T) \sigma^{-1}\,, \tag{1.5}$$

$$n^{\lambda'} = \sum_{\sigma \in \mathcal{S}_n} \sigma N(T) \sigma^{-1}\,, \tag{1.6}$$

$$\chi^\lambda = \sum_{\sigma \in \mathcal{S}_n} \sigma Y(T) \sigma^{-1}\,. \tag{1.7}$$

Next we consider the space Λ^n. Let $x_1, \dots, x_N$ be any set of variables where $N \geq n$. A polynomial $P(x_1, \dots, x_N)$ over $\mathbb{C}$ is *symmetric* if and only if

$$P(x_1, \dots, x_N) = P(x_{\sigma_1}, \dots, x_{\sigma_N})$$

for all $\sigma \in \mathcal{S}_N$. We let $\Lambda^n = \Lambda^n(x_1, \dots, x_N)$ denote the space of homogeneous symmetric polynomials of degree n. There are 5 fundamental bases for Λ^n. Given a monomial $m = x_1^{i_1} \cdots x_N^{i_N}$ of degree n, we let the type of m, $\tau(m)$, be the partition of n induced by increasing rearrangement of the exponents $(i_1, \dots, i_N)$. For example, $\tau(x_1^2 x_3^3 x_5 x_6^2) = (1, 2, 2, 3)$. Then for any $\lambda \vdash n$, the *monomial symmetric function* m_λ is defined by

$$m_\lambda = \sum_{i_1, \dots, i_N} x_1^{i_1} \cdots x_N^{i_N} \chi\left(\tau\left(x_1^{i_1} \cdots x_N^{i_N}\right) = \lambda\right)\,. \tag{1.8}$$

Since permuting the indices of a monomial preserves its type, it easily follows that $\{m_\lambda\}_{\lambda \vdash n}$ forms a basis for Λ^n. Thus the dimension of Λ^n and $Z(\mathcal{S}_n)$ are both equal to the number of partitions of n.

Now there are three so-called multiplicative bases of Λ^n. Namely, given $\lambda = (0 < \lambda_1 \leq \cdots \leq \lambda_k)$, we define the *homogeneous symmetric function* h_λ by

$$h_\lambda = h_{\lambda_1} h_{\lambda_2} \cdots h_{\lambda_k} \tag{1.9}$$

where for each integer r,

$$h_r = \sum_{1 \leq i_1 \leq \cdots \leq i_r \leq N} x_{i_1} \cdots x_{i_r}\,.$$

The *elementary symmetric function* e_λ is defined by

$$e_\lambda = e_{\lambda_1} e_{\lambda_2} \cdots e_{\lambda_k} \tag{1.10}$$

where for each integer r,

$$e_r = \sum_{1 \le i_1 < \cdots < i_r \le N} x_{i_1} \cdots x_{i_r} .$$

We define the *power symmetric function* ψ_λ by

$$\psi_\lambda = \psi_{\lambda_1} \psi_{\lambda_2} \cdots \psi_{\lambda_k} \tag{1.11}$$

where for each integer r,

$$\psi_r = \sum_{j=1}^{N} x_j^r .$$

Our final basis is the Schur functions. We can define a Schur function S_λ in two different ways. In terms of determinants we have

$$S_\lambda = \frac{\det \|x_i^{\lambda_j + n - j}\|}{\det \|x_i^{n-j}\|} . \tag{1.12}$$

A more combinatorial definition of S_λ is the following. Given a column strict tableau T with entries from the set $\{1, \ldots, N\}$, define the monomial weight $m(T)$ of T by $m(T) = x_1^{i_1} \cdots x_N^{i_N}$ where for each j, i_j denotes the number of times the integer j occurs in T. For example, for the column strict tableau T_1 given earlier in Figure 1.1, $m(T_1) = x_1^2 x_2^2 x_3 x_4^2 x_5$. If $\mathcal{T}_\lambda$ denotes the set of all column strict tableaux T of shape λ with entries from $\{1, \ldots, N\}$, an alternative definition of S_λ is

$$S_\lambda = \sum_{T \in \mathcal{T}_\lambda} m(T) . \tag{1.13}$$

By a vertical bar, we denote the operation of taking coefficients. Thus for example $\psi_\lambda\big|_{h_\mu}$ denotes the coefficient of h_μ in the expansion of ψ_λ in the homogeneous basis.

§2. The Frobenius map

The Frobenius map $F: Z(\mathcal{S}_n) \to \Lambda^n$ is defined as follows.

$$F(f) = \frac{1}{n!} \sum_{\sigma \in \mathcal{S}_n} f(\sigma)\psi_{\lambda(\sigma)} \,. \tag{2.1}$$

The Frobenius map is an isometry from $Z(\mathcal{S}_n)$ onto Λ^n where the scalar product on Λ^n is the so-called Hall scalar product defined by declaring $\langle h_\lambda, m_\mu \rangle = \chi(\lambda = \mu)$. In terms of the various bases for $Z(\mathcal{S}_n)$ and Λ^n defined in section 2, the following hold

$$F(z_\lambda C^\lambda) = \psi_\lambda \tag{2.2}$$

$$F(p^\lambda) = h_\lambda \tag{2.3}$$

$$F(n^\lambda) = e_\lambda \tag{2.4}$$

$$F(\chi^\lambda) = S_\lambda \,. \tag{2.5}$$

A complete proof may be found in [**13**].

Next we introduce three fundamental products of Schur functions and describe their connections with the representation theory of $\mathcal{S}_n$. Let A_λ and A_μ denote irreducible representation corresponding to the partitions $\lambda \vdash n$ and $\mu \vdash m$. The ordinary product of Schur functions corresponds to inducing $A_\lambda \times A_\mu$ from $\mathcal{S}_n \times \mathcal{S}_m$ to $\mathcal{S}_{n+m}$. That is,

$$S_\lambda S_\mu = F\left(\chi^{ind(A_\lambda \times A_\mu \uparrow^{\mathcal{S}_{n+m}}_{\mathcal{S}_n \times \mathcal{S}_m}}\right) \,. \tag{2.6}$$

The Kronecker product of Schur functions, denoted by $S_\lambda \otimes S_\mu$, corresponds to the tensor product of the representation $A_\lambda \otimes A_\mu$. Here we must assume $m = n$. That is,

$$S_\lambda \otimes S_\mu = F\left(\chi^{A_\lambda \otimes A_\mu}\right) \,. \tag{2.7}$$

Because the Schur functions are a basis for Λ^n, we can extend our definition of $\otimes$ to all Λ^n via (2.7) by linearity.

Next we consider the definition of the plethsym $S_\lambda[S_\mu]$ of two Schur functions. The plethsym of two symmetric polynomials has a more complex definition than the ordinary product or the Kronecker product of Schur functions. Given symmetric polynomials with integer coefficients, $P \in \Lambda^n$ and $Q \in \Lambda^m$, $P[Q]$ will be a symmetric polynomial in $\Lambda^{n \cdot m}$ defined as follows. First write $Q = \sum_\alpha a_\alpha x^\alpha$ where a_α is an integer and if $\alpha = (\alpha_1, \alpha_2, \dots)$, then $x^\alpha = x_1^{\alpha_1} x_2^{\alpha_2} \cdots$. Then define

$$e_r^Q(x) = \prod_\alpha (1 + t\, x^\alpha)^{a_\alpha}|_{t^r} \tag{2.8}$$

and

$$e_\lambda^Q(x) = e_{\lambda_1}^Q(x) \cdots e_{\lambda_k}^Q(x) \quad \text{if} \quad \lambda = (\lambda_1, \dots, \lambda_k)\,. \tag{2.9}$$

Since P is a symmetric polynomial, we can express P in terms of the elementary symmetric functions,

$$P = \sum_{\lambda \vdash n} c_\lambda\, e_\lambda(x)\,. \tag{2.10}$$

Then by definition,

$$P[Q] = \sum_{\lambda \vdash n} c_\lambda\, e_\lambda^Q(x)\,. \tag{2.11}$$

It is worth noting that if $Q = \sum_\alpha a_\alpha\, x^\alpha$ and a_α is nonnegative for all α, then $P[Q]$ can be described by a kind of substitution operation. This is best illustrated by an example. Suppose we want to compute $S_{2,3}[S_2](x_1, x_2, x_3)$. First we express $S_2(x_1, x_2, x_3)$ as a sum of monomials.

$$S_2(x_1, x_2, x_3) = x_1^2 + x_1x_2 + x_1x_3 + x_2^2 + x_2x_3 + x_3^2\,. \tag{2.12}$$

Then we think of $S_{2,3}$ as a function of as many variables as there are monomials in $S_2(x_1, x_2, x_3)$. In this case, there are six monomials in $S_2(x_1, x_2, x_3)$ so we consider $S_{(2,3)}(x_1, x_2, \dots, x_6)$. Then $S_{(2,3)}[S_2](x_1, x_2, x_3)$ results by substituting the monomials occuring in $S_2(x_1, x_2, x_3)$ for the variables in $S_{(2,3)}(x_1, \dots, x_6)$. That is,

$$S_{(2,3)}[S_2](x_1, x_2, x_3) = S_{(2,3)}(x_1^2,\ x_1x_2,\ x_1x_3,\ x_2^2,\ x_2x_3,\ x_3^2)\,. \tag{2.13}$$

As with the ordinary product and Kronecker product, the plethysm of two Schur functions is the Frobenius image of a character of the symmetric group. Again the situation is more complicated than is the case for either the ordinary product or Kronecker product. Without going into details, the idea is to start out with an irreducible $\mathcal{S}_n$-module U_λ corresponding to the partition $\lambda \vdash n$ and an irreducible $\mathcal{S}_m$-module corresponding to the partition $\mu \vdash m$. Let $U_\mu^{\otimes n}$ denote the n-fold tensor product of U_μ. Then the wreath product of $\mathcal{S}_n$ with $\mathcal{S}_n$, $\mathcal{S}_n(\mathcal{S}_m)$, acts naturally on $U_\lambda \otimes U_\mu^{\otimes n}$ and

$$S_\lambda[S_\mu] = F\left(\chi^{ind_{\mathcal{S}_n(\mathcal{S}_m)}^{\mathcal{S}_{n\cdot m}}(U_\lambda \otimes U_\mu^{\otimes n})}\right). \tag{2.14}$$

See [**22**] or [**16**] for details.

We note that since the Schur functions are also the characters of the irreducible polynomial representations of $GL_n(\mathbb{C})$, the ordinary product and the plethysm of Schur functions are easily interpreted as characters of representations of $GL_n(\mathbb{C})$. That is, the ordinary product of Schur functions, $S_\lambda S_\mu$, is just the character of the tensor product of the irreducible representation of the general linear group which correspond to the characters S_λ and S_μ respectively. Similarly $S_\lambda[S_\mu]$ can be thought of as the character which arises from composing the irreducible representations with character S_λ with the irreducible representation with character S_μ. See [**15**] or [**37**] for details.

§3. Combinatorial methods for expanding $S_\lambda S_\mu$ and $\psi_n S_\mu$

In this section we shall describe how to expand $S_\lambda S_\mu$ and $\psi_n S_\mu$ as a product of Schur functions. First we need some notation.

Given two partitions $I = (i_1, \ldots, i_k)$ and $J = (j_1, \ldots, j_\ell)$, we write $I \leq J$, if and only if $k \leq \ell$ and $i_{k-p} \leq j_{\ell-p}$ for $p = 0, \ldots, k-1$. If $I \leq J$, we let $F_{J/I}$ denote the Ferrers' diagram of the skew shape J/I where $F_{J/I}$ is the diagram that results by removing the cells corresponding to F_I from the diagram F_J. For example

$F_{(1,2,3,3)/(1,2)} =$

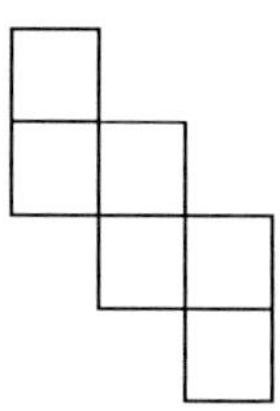

A column strict tableau T of shape J/I is a filling of $F_{J/I}$ with positive integers so that the numbers weakly increase from left to right in each row and strictly increase from bottom to top in each column. T is said to be *standard* if the entries of T are precisely $1, \ldots, n$ where $n = |J/I| = \#$ of cells in $F_{J/I}$. We let $\mathcal{T}_{J/I}$ and $\mathcal{ST}_{J/I}$ denote the set of all column strict tableaux and standard tableaux of shape J/I respectively. Given $T \in \mathcal{T}_{J/I}$, the weight of $T, m(T)$, is the monomial that results by replacing each cell with an i in it by x_i and taking the product over all cells. For example, if

$T =$ 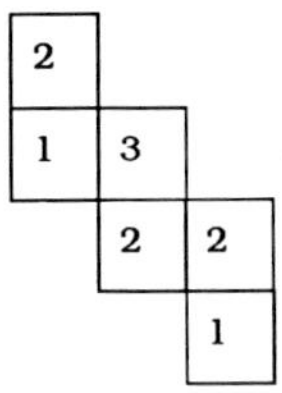 , then $m(T) = x_1^2 x_2^3 x_3$.

This given, the skew Schur function $S_{J/I}$ is defined by

(3.1) $$S_{J/I}(x_1 x_2, \ldots) = \sum_{T \in \mathcal{T}_{J/I}} m(T) .$$

The special case of (3.1) where I is the empty diagram, i.e. $I = \emptyset$, defines the usual Schur function S_J. Given $T \in \mathcal{T}_{J/I}$, we shall write $sh(T)$ for the shape of T, i.e. $sh(T) = J/I$. For emphasis, we shall often refer to those shapes which arise directly from partitions J as straight shapes so as to distinguish them among the general class of skew shapes. If n is an integer, we shall write S_n for the Schur function corresponding to the partition of n consisting of one part of size n.

Given a skew shape or Ferrers' diagram D, we shall say $D_1 + \cdots + D_k = D$ is a *decomposition* of D if there exists an increasing sequence of shapes $I = I_0 \leq I_1 \leq \cdots \leq I_k = J$ such that $D = J/I$ and $D_i = I_i/I_{i-1}$ for $i = 1, \ldots, k$. For example, $(1,2) + (1,2,3,3)/(1,2) + (1,3,4,4,4)/(1,2,3,3)$ is a decomposition of the shape $(1,3,4,4,4)$ corresponding to the increasing sequence of diagrams $\emptyset \leq (1,2) \leq (1,2,3,3) \leq (1,3,4,4,4)$.

Next we need to consider rules to compute the so-called Littlewood-Richardson coefficients which are defined by

(3.2) $$g_{IJK} = \langle S_I S_J, S_K \rangle .$$

Let $I * J$ denote the skew diagram that results from F_I and F_J by placing F_I on top of F_J so that the start of the top row of F_J is just below the end of the bottom row of F_I. For example,

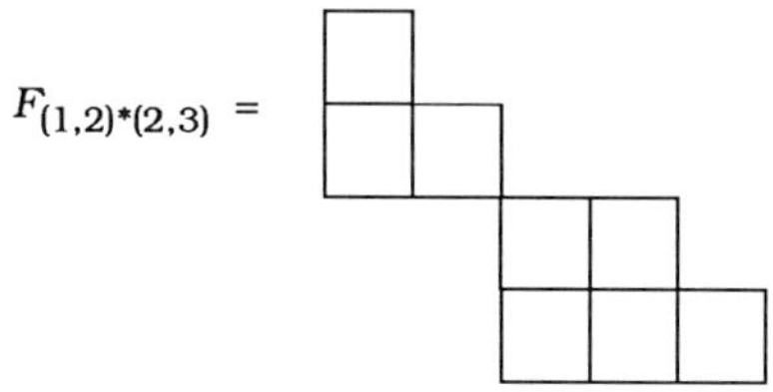

Then clearly $S_{I*J} = S_I S_J$ so that computing the g_{IJK}'s is just a special case of expanding an arbitrary skew Schur function as a sum of Schur functions. Moreover it should be clear that the problem of expanding an arbitrary product of Schur functions or skew Schur functions corresponds to expanding a single skew Schur function as a sum of Schur functions. For example, $S_{(2,3)} \cdot S_{(1,2)} \cdot S_{(4,4,4)/(1,2)}$ is equal to the skew Schur function whose Ferrers' diagram is pictured below.

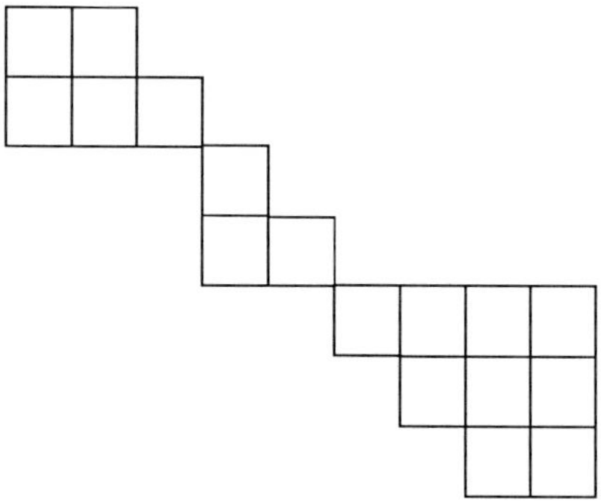

Such expansions can be computed via the following version of the Littlewood-Richardson rule due to Remmel and Whitney [**30**].

Skew Schur Function Expansion Rule

To compute $S_{J/I} = \sum_K g_{J/I,K} S_K$.

(1) Form the *reverse lexicographic filling* of J/I, $r\ell(J/I)$, which is the filling of $F_{J/I}$

which starts at the bottom right corner of $F_{J/I}$ and fills in the integers $1, 2, \ldots, n$ in order from right to left and bottom to top. For example

$$r\ell(F_{(1,2,3,3)/(1,2)}) = \begin{array}{|c|c|c|} \hline 6 & & \\ \hline 5 & 4 & \\ \hline & 3 & 2 \\ \hline & & 1 \\ \hline \end{array} \quad .$$

(2) We say a standard tableau T is (J/I)-compatible if

(a) whenever $i+1$ is immediately to the left of i in $r\ell(J/I)$, then in T, $i+1$ occurs to the southeast of i in the sense that the cell of T which contains $i+1$ is strictly to the right and weakly below the cell of T which contains i.

(b) whenever y is immediately above x in $r\ell(J/I)$, then in T, y occurs to the northwest of x in the sense that the cell of T which contains y is strictly above and weakly to the left of the cell of T which contains x.

Then $g_{J/I,K}$ is the number of (J/I)-compatible tableaux of shape K.

It is good to visualize the condition '$i+1$ southeast of i' and 'y northwest of x' respectively by the patterns

$$i_{i+1} \quad y_x \ .$$

Thus for our example above, conditions (a) and (b) may be summarized by the patterns

$$2_3 \ , \ 4_5 \ ,$$

$$2_1 \ , \ 4_3 \ , \ 6_5 \ .$$

This given, the collection of (J/I)-compatible tableaux can easily be constructed by adding squares labeled $1, 2, \ldots, n$ in succession, always maintaining standardness and obeying each time the conditions (a) and (b). In our example, one is naturally led to the following tree for constructing the (J/I)-compatible tableaux.

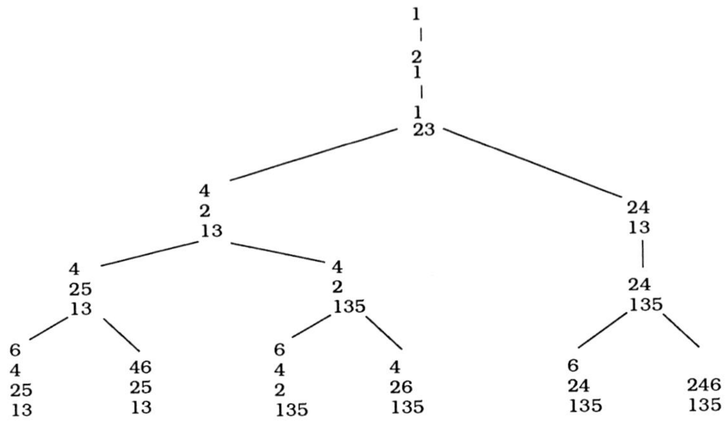

Having constructed the tree, one can easily read off the expansion of $S_{J/I}$ as

(3.3) $$S_{J/I} = \sum_{\substack{T \\ T(J/I)\text{-compatible}}} S_{sh(T)}$$

where $sh(T)$ denotes the shape of T. Thus in our example,

$$S_{(1,2,3,3)/(1,2)} = S_{(1,1,2,2)} + S_{(2^3)} + S_{(1,1,1,3)} + 2S_{(1,2,3)} + S_{(3,3)}\,.$$

The importance of being able to calculate $\psi_n S_\mu$ is that it allows one to compute the value χ^λ_μ of the irreducible character χ^λ at the conjugacy class C^μ. That is, it is easy to see via the Frobenius map that

(3.4) $$\psi_\mu = \sum_{\lambda \vdash n} \chi^\lambda_\mu S_\lambda\,.$$

To give a combinatorial method to expand $\psi_n S_\mu$ or compute χ^λ_μ, we need the concept of a *rim hook*. A rim hook H of a partition λ is a consecutive sequence of cells on the northeastern rim of F_λ such that any two adjacent cells of H share a common edge and the removal of H from F_λ leaves of Ferrers' diagram corresponding to a partition. We let $r(H)$ denote the number of rows of H. Then the sign of H, $sgn(H)$, is defined by

$$sgn(H) = (-1)^{r(H)-1}\,.$$

For example below is pictured all rim hooks of size 4 for $\lambda = (2,3,3,4)$.

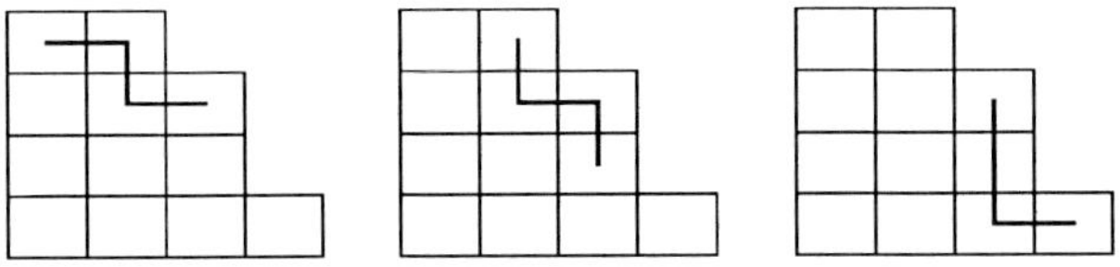

The respective signs of the rim hooks from left to right are $(-1)^{2-1}$, $(-1)^{3-1}$, and $(-1)^{3-1}$. This given,

$$\psi_n S_\mu = \sum_{\substack{\lambda \geq \mu \\ \lambda \vdash n+|\mu| \\ \lambda/\mu \text{ is a rim hook of size } n}} sgn(\lambda/\mu) S_\lambda \,. \tag{3.5}$$

Thus the shapes λ for which S_λ appears expansion of $\psi_n S_\mu$ are gotten by placing a rim hook of size n along the outside of F_μ in all possible ways. For example, placing a rim hook of size 3 along the outside of $F_{(2,3)}$ leads to the following shapes

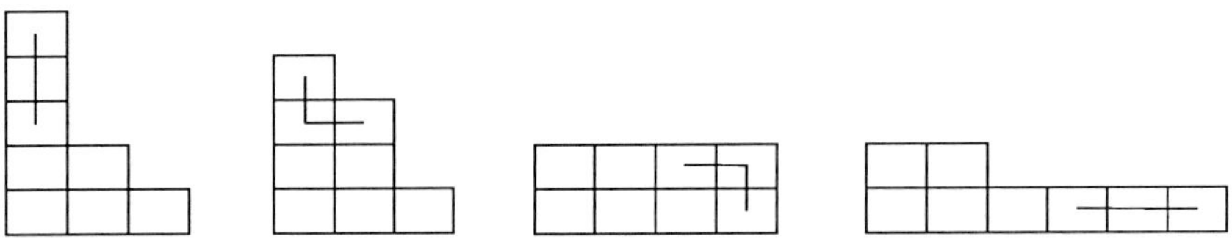

We can then read off the expansion of $\psi_3 S_{(2,3)}$ as

$$\psi_3 S_{(2,3)} = S_{(1^3,2,3)} - S_{(1,2^2,3)} - S_{(4,4)} + S_{(2,6)} \,.$$

Now we can expand ψ_μ for any partition μ in terms of Schur functions by repeated use of (3.5). This leads to the Murnagham-Nakayama rule for the computation of χ^λ_μ. A rim hook tableau of shape λ and type $\mu = (\mu_1 \leq \mu_2 \leq \cdots \leq \mu_k)$ is a filling T of F_λ with rim hooks $H_1, \ldots, H_k$ of sizes $\mu_1, \ldots, \mu_k$ respectively such that if $\lambda^{(i)}$ denotes the set of cells filled by rim hooks $H_1, \ldots, H_i$, then $\lambda^{(i)}$ is a partition and H_i is a rim hook of $\lambda^{(i)}$. The sign of T, $sgn(T)$, is given by

$$sgn(T) = \prod_{i=1}^{k} sgn(H_i) \,. \tag{3.6}$$

Then

$$\chi_\mu^\lambda = \sum_T sgn(T) \tag{3.7}$$

where the sum runs over all rim hook tableaux of shape λ and type μ. For example, to compute χ_μ^λ where $\lambda = (3,4)$ and $\mu = (2,2,3)$, we must fill F_λ with rim hooks H_1, H_2, and H_3 where $|H_1| = 2$, $|H_2| = 2$, and $|H_3| = 3$. It is then easy to check that if we specify the cells of H_i by placing i in each of the cells, then there are 3 rim hook tableaux of shape λ and type μ.

3	3	3	
1	1	2	2

T_1

2	2	3	
1	1	3	3

T_2

1	2	3	
1	2	3	3

T_3

Thus

$$\begin{aligned} \chi_{(2,2,3)}^{(3,4)} &= sgn(T_1) + sgn(T_2) + sgn(T_3) \\ &= 1 + (-1) + (-1) = -1\,. \end{aligned}$$

§4. Combinatorial methods for the expansion of $S_\lambda \otimes S_\mu$

In this section we present a combinatorial algorithm for the expansion of $S_\lambda \otimes S_\mu$ which was given by Garsia and Remmel in [**12**]. Recall that if λ and μ are partitions of n and A^λ and A^μ are irreducible representations of $\mathcal{S}_n$, then

$$S_\lambda \otimes S_\mu = F\left(\chi^{A_\lambda \otimes A_\mu}\right) = F(\chi^\lambda \otimes \chi^\mu) \tag{4.1}$$

where $\chi^\lambda = \chi^{A_\lambda}$, $\chi^\mu = \chi^{A_\mu}$, and for all $\sigma \in \mathcal{S}_n$, $\chi^\lambda \otimes \chi^\mu(\sigma) = \chi^\lambda(\sigma) \cdot \chi^\mu(\sigma)$. Thus the Kronecker product of Schur functions corresponds to the pointwise product in $Z(\mathcal{S}_n)$. This fact allows one to easily establish the following:

$$S_n \otimes S_\lambda = S_\lambda\,, \tag{4.2}$$

$$S_{1^n} \otimes S_\lambda = S_{\lambda'}\,, \tag{4.3}$$

$$S_\lambda \otimes S_\beta = S_\beta \otimes S_\lambda = S_{\lambda'} \otimes S_{\beta'} = S_{\beta'} \otimes S_{\lambda'} \,. \tag{4.4}$$

In fact, if we let $S_\lambda \otimes S_\mu = \sum_\nu c_{\nu,\lambda,\mu} S_\nu$, then

$$\begin{aligned} c_{\nu,\lambda,\mu} &= \langle S_\lambda \otimes S_\mu, S_\nu \rangle \\ &= \langle \chi^\lambda \otimes \chi^\mu, \chi^\nu \rangle \\ &= \frac{1}{n!} \sum_{\sigma \in \mathcal{S}_n} \chi^\lambda(\sigma)\chi^\mu(\sigma)\overline{\chi^\nu(\sigma)} \end{aligned} \tag{4.5}$$

It is easy to see from section 3 that the characters χ^ν are real valued, hence it follows from (4.4) that $c_{\nu,\lambda,\mu} = \frac{1}{n!} \sum_{\sigma \in \mathcal{S}_n} \chi^\lambda(\sigma)\chi^\mu(\sigma)\chi^\mu(\sigma)$. Thus $c_{\nu,\lambda,\mu}$ is symmetric in λ, μ, and ν. Littlewood [**21**] proved the following

$$(S_\rho S_\tau) \otimes S_\nu = \sum_{\substack{\lambda \vdash |\rho| \\ \mu \vdash |\tau|}} g_{\lambda\mu\nu} (S_\lambda \otimes S_\rho)(S_\mu \otimes S_\tau) \tag{4.6}$$

where ρ, τ, λ, μ, are straight shapes and $g_{\lambda\mu\nu}$ is the Littlewood-Richardson coefficients as defined in section 3. Garsia and Remmel [**12**] then used (4.6) to prove the following:

$$(S_H,\, S_K) \otimes S_D = \sum_{\substack{D_1+D_2=D \\ |D_1|=|H| \\ |D_2|=|K|}} (S_H \otimes S_{D_1}),\ (S_K \otimes S_{D_2}) \tag{4.7}$$

where here H, K, and D can be arbitrary skew shapes and the sum runs over all decompositions of the skew shape D. In particular, one can easily establish by induction from (4.7) that

$$(S_{n_1} \cdots S_{n_k}) \otimes S_D = \sum_{\substack{D_1+\cdots+D_k=D \\ |D_i|=n_i}} S_{D_1} \cdots S_{D_k} \tag{4.8}$$

where the sum runs over all decompositions of D of length k such that $|D_i| = n_i$ for all i.

Then (4.8) allows us to compute $S_I \otimes S_J$ by using the Jacobi-Trudi identity (see [**22**]).

$$S_{\lambda/\mu} = \det |S_{(\lambda_i+i)-(\mu_j+j)}| \tag{4.9}$$

where $\lambda = (0 \le \lambda_1 \le \cdots \le \lambda_n)$, $\mu = (0 \le \mu_1 \le \cdots \le \mu_n)$, and $\mathcal{S}_n$ is interpreted to be zero if n is negative.

As an example consider the computation of $S_{(3,3)} \otimes S_{(1^3,3)}$. First we express

$$S_{(3,3)} = \det \begin{vmatrix} S_3 & S_4 \\ S_2 & S_3 \end{vmatrix} = S_3S_3 - S_2S_4$$

via the Jacobi-Trudi identity. Then by (4.8)

(4.10) $$S_3S_3 \otimes S_{(1^3,3)} = S_3S_{(1^3,3)/3} + S_{(1,2)}S_{(1^3,3)/(1,2)} + S_{(1^3)}S_{(1^3,3)/(1^3)}$$

= S 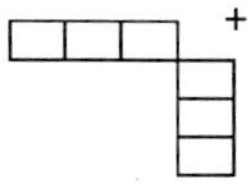 + S 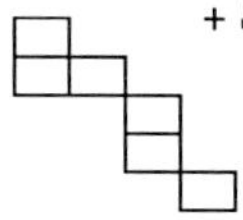 + S 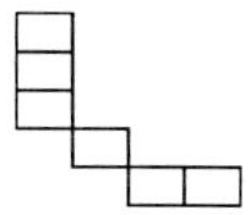

and

(4.11) $$-S_{(2,4)} \otimes S_{(1^3,3)} = -S_2S_{(1^3,3)/2} - S_{1^2}S_{(1^3,3)/(1^2)}$$

= - S 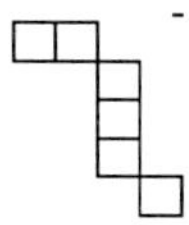 - S 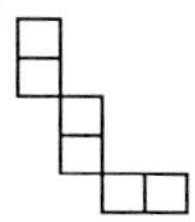

Finally we can easily compute (4.10) and (4.11) by our skew Schur function expansion rule. That is, one can easily calculate that

S 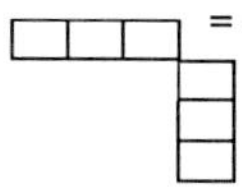 $= S_{(1^3,3)} + S_{(1^2,4)}$,

S 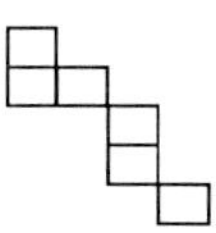 $= S_{(1^4,2)} + 2S_{(1^3,3)} + 2S_{(1^2,2^2)} + 3S_{(1,2,3)} + S_{(1^2,4)} + S_{(2^3)} + S_{(3,3)} + S_{(2,4)}$,

S 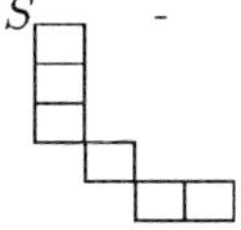 - S 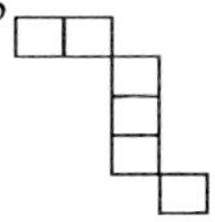 $= 0,$

$-S$ 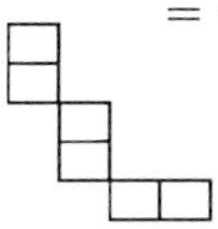 $= -S_{(1^4,2)} - S_{(1^2,2^2)} - 2S_{(1^3,3)} - 2S_{(1,2,3)} - S_{(1^2,4)} - S_{(2^3)} - S_{(2,4)}.$

Thus

$$S_{(3,3)} \otimes S_{(1^3,3)} = S_{(1^2,2^2)} + S_{(1^3,3)} + S_{(1,2,3)} + S_{(1^2,4)} + S_{(3,3)}\,.$$

Note that the ability to efficiently compute arbitrary products of skew Schur functions is crucial to carry out the above algorithm. But nevertheless, the algorithm is nonefficient due to the cancellation that arises due to the determinant. There is nothing like the Littlewood-Richardson rule to directly calculate $c_{\lambda,\mu,\nu}$. We should note however that Remmel [**26,27**] used this algorithm plus some combinatorial involutions to deal with the cancellations to compute explicit formulas for $S_\lambda \otimes S_\mu$ where λ is hook shape, i.e. λ is of the form $(1^r, n-r)$, and μ is either a hook shape, a partition with ≤ 2 parts, or a partition with largest part ≤ 2. In such cases, $c_{\lambda,\mu,\nu}$ is always either $0, 1, 2$, or 3 which shows that there is a tremendous amount of cancellation which goes on in the above algorithm.

§5. Combinatorial methods for expanding $S_\lambda[S_\mu]$

In this section we present some combinatorial algorithms to compute $S_\lambda[S_\mu]$ which can be found in [**3**]. We start by listing several basic properties of plethysms. Let P_1, P_2, and Q be symmetric polynomials. Then

(5.1) $$P_1 \pm P_2[Q] = P_1[Q] \pm P_2[Q]$$

(5.2) $$P_1 \cdot P_2[Q] = P_1[Q]P_2[Q]$$

(5.3) $$\psi_n[Q] = Q[\psi_n] = Q(x^p)$$

where $x^p = (x_1^p, \ldots, x_N^p)$. Note that (5.1)-(5.3) allow us to show that we can easily calculate plethysms with respect to the basis of power symmetric functions. That is, if $\lambda = (\lambda_1, \ldots, \lambda_k)$ and $\mu = (\mu_1, \ldots, \mu_\ell)$ are partitions, then we let $\lambda * \mu$ denote the partition of $|\lambda| \cdot |\mu|$ whose parts are $\lambda_i \mu_j$ for $i = 1, \ldots, k$ and $j = 1, \ldots, \ell$. Then

(5.4) $$\psi_\lambda[\psi_\mu] = \psi_\mu[\psi_\lambda] = \psi_{\lambda * \mu} .$$

In addition, we have the following formulas.

(5.6) $$S_n[P \cdot Q] = \sum_{\lambda \vdash n} S_\lambda[P]\, S_\lambda[Q]$$

(5.7) $$S_\lambda[P + Q] = \sum_{\mu \subseteq \lambda} S_\mu[P]\, S_{\lambda/\mu}[Q]$$

(5.8) $$(S_\lambda[S_\mu])' = \begin{cases} S_\lambda[S_{\mu'}] & \text{if } |\mu| \text{ is even} \\ S_{\lambda'}[S_{\mu'}] & \text{if } |\mu| \text{ is odd} \end{cases}$$

where for any sum $\sum c_\nu S_\nu$, $(\sum c_\nu S_\nu)'$ denotes the sum $\sum c_\nu S_{\nu'}$.

The main problem is to compute $a^\nu_{\lambda,\mu}$ where

(5.9) $$S_\lambda[S_\mu] = \sum_{\nu \vdash |\lambda| \cdot |\mu|} a^\nu_{\lambda,\mu} S_\nu .$$

Since $S_\lambda[S_\mu]$ is the Frobenius image of a character, see (2.14), we know $a^\nu_{\lambda,\mu}$ are nonnegative integers. There are basically two approaches to the calculation of $S_\lambda[S_\mu]$. The first is to expand S_λ in terms of the power symmetric functions,

(5.10) $$S_\lambda = \sum_{p_1 + 2p_2 + \cdots + np_n = n} \frac{\chi^\lambda_{(1^{p_1} 2^{p_2} \cdots n^{p_n})}}{p_1! \cdots p_n!} \left(\frac{\psi_1}{1}\right)^{p_1} \cdots \left(\frac{\psi_n}{n}\right)^{p_n} .$$

Then using (5.1) - (5.3), we see that

(5.11)
$$S_\lambda[S_\mu] = \sum_{p_1 + 2p_2 + \cdots + np_n = n} \frac{\chi^\lambda_{(1^{p_1} 2^{p_2} \cdots n^{p_n})}}{p_1! \cdots p_n!} \left(\frac{S_\mu(x)}{1}\right)^{p_1} \left(\frac{S_\mu(x^2)}{2}\right)^{p_2} \cdots \left(\frac{S_\mu(x^n)}{n}\right)^{p_n} .$$

On the other hand, using the Jacobi-Trudi identity, (5.1), and (5.2), we see that

$$S_\lambda[S_\mu] = \det |S_{\lambda_i+i-j}[S_\mu]| \,. \tag{5.12}$$

For example, when $\lambda = (1,3)$, (5.11) reduces to

$$\begin{aligned} S_{(1,3)}[S_\mu] = & \frac{1}{24}\chi^{(1,3)}_{(1^4)}\,(S_\mu(x))^4 + \frac{1}{4}\chi^{(1,3)}_{(1^2,2)}\,S_\mu(x)^2 S_\mu(x^2) \\ & + \frac{1}{8}\chi^{(1,3)}_{(2,2)}\,(S_\mu(x^2))^2 + \frac{1}{3}\chi^{(1,3)}_{(1,3)}\,S_\mu(x)S_\mu(x^3) \\ & + \frac{1}{4}\chi^{(1,3)}_{(4)}\,S_\mu(x^4) \end{aligned} \tag{5.13}$$

while (5.12) reduces to

$$\begin{aligned} S_{(1,3)}[S_\mu] &= \det \begin{vmatrix} S_1[S_\mu] & S_4[S_\mu] \\ 1 & S_3[S_\mu] \end{vmatrix} \\ &= S_\mu(x)S_3[S_\mu] - S_4[S_\mu]\,. \end{aligned} \tag{5.14}$$

To compute $S_n[S_\mu]$, we can use the fact that

$$S_n = \frac{1}{n}\sum_{p=1}^{n} \psi_p S_{n-p} \tag{5.15}$$

to give us the recursion

$$S_n[S_\mu] = \frac{1}{n}\sum_{p=1}^{n} S_\mu(x^p) S_{n-p}[S_\mu]\,. \tag{5.16}$$

At any rate, which ever method we choose, we are reduced to expanding arbitrary products of Schur functions, which can be done via the algorithm in section 3, and calculations of Schur function expansions for the plethysms $S_\mu(x^p)$.

We next describe a general algorithm to compute $S_\mu(x^p)$ for an arbitrary μ and then a more efficient algorithm to compute $S_n(x^p)$.

Let μ be a partition of n and let us suppose we are given to calculate the expansion of $S_\mu(x^p)$ for some $p \geq 2$. The point of departure is the formula

$$S_\mu(x^p) = \sum_{|I_1|+\cdots+|I_p|=n} c_{I_1,\ldots,I_p} SS_{I_1,\ldots,I_p}(x) \tag{5.17}$$

where

(a) the sum is to be carried out over all p-tuples of partitions $I_1, \ldots, I_p$ whose diagrams are contained in μ and whose sum of parts add up to n,

(b) we have

$$c_{I_1,\ldots,I_p} = S_{I_1} S_{I_2} \cdots S_{I_p}|_{S_\mu} \tag{5.18}$$

and

(c) the expression $SS_{I_1,\ldots,I_p}(x)$ denotes a certain signed Schur function indexed by a partition with empty p-core whose construction is best explained through an example.

For instance, in the expansion of $S_{113}(x^3)$, since $n = 5$ and $p = 3$, one of the terms in (5.17) is that which corresponds to the triple of partitions

$$I_1 = \square, I_2 = \begin{smallmatrix}\square\\ \square\end{smallmatrix}, I_3 = \square\square .$$

The Schur function multiplication algorithm based on the diagram

$$D = \begin{array}{cccc} 5 & & & \\ & 4 & & \\ & 3 & & \\ & & 2 & 1 \end{array}$$

produces two D-compatible tableaux of shape 113, namely

$$\begin{array}{lll} 5 & & \\ 4 & & \\ 1 & 2 & 3 \end{array} \quad \text{and} \quad \begin{array}{lll} 4 & & \\ 3 & & \\ 1 & 2 & 5 . \end{array}$$

This gives that

$$c_{\square,\, \begin{smallmatrix}\square\\ \square\end{smallmatrix},\, \square\square}(113) = 2 . \tag{5.19}$$

To construct

$$SS_{\square,\, \begin{smallmatrix}\square\\ \square\end{smallmatrix},\, \square\square}(x) ,$$

we proceed as follows. First of all we represent $\square$, $\begin{smallmatrix}\square\\\square\end{smallmatrix}$, and $\square\square$ as partitions with an equal number of parts, that is, we set

$$\square = (0,1),\ \begin{smallmatrix}\square\\\square\end{smallmatrix} = (1,1),\ \square\square = (0,2)\,.$$

This given we construct the *circle diagram* given below

(5.20)

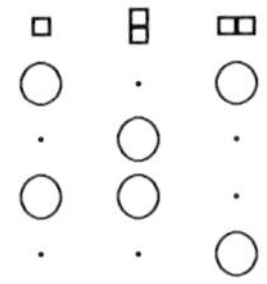

The precise rule for putting together a column of this diagram from a partition $(i_1, i_2, \ldots, i_m)$ is that the distance (in dots) between the sth and $(s+1)^x$ circle is given by the difference $i_{s+1} - i_s$, or (which is the same) the sth circle is at distance $i_s + s - 1$ from the top. That is pictorially we have

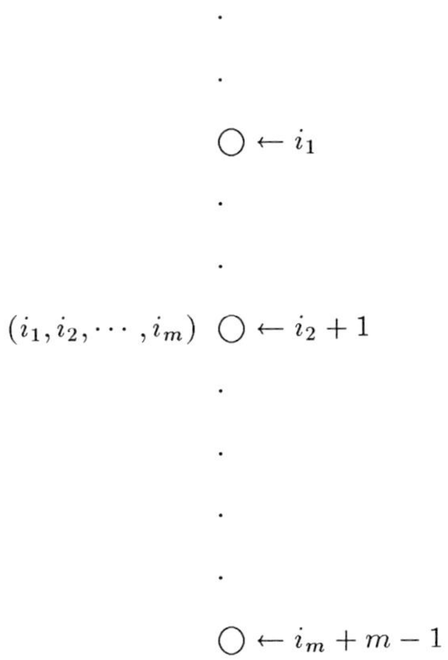

Accordingly, in the column labelled by $\begin{smallmatrix}\square\\\square\end{smallmatrix} = (1,1)$ the first circle is at distance 1 from the top and the second circle is at distance one from the first. Proceeding in the same manner for the other two partitions we obtain the circle diagram given in (5.20). This done, we assign to the positions in the diagram (indicated by dots when not by circles) the labels $0, 1, 2, 3, \ldots$ successively from left to right and from top to bottom, and record the label only when it falls in one of the circles. This gives the labelled diagram

(5.21)

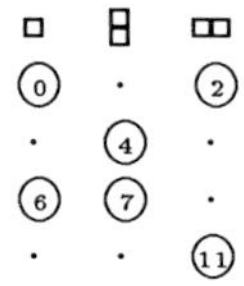

In the case of a general p-tuple $I_1, I_2, \ldots, I_p$ we obtain a circle diagram with m circles in each column, where m is the maximum number of non-zero parts appearing in any of the partitions $I_1, I_2, \ldots, I_p$.

Let $b_1 < b_2 < b_3 < \cdots < b_{mp}$ be the labels placed on the circles and $q_{s,1} < q_{s,2} < \cdots < q_{s,m}$ be the labels appearing in the column corresponding to the partition I_s. Finally, let $inv(I_1, \ldots, I_p)$ denote the number of inversions of the permutation

$$q_{1,1}\ q_{1,2} \cdots q_{1,m}\ q_{2,1}\ q_{2,2} \cdots q_{2,m} \cdots q_{p,1}\ q_{p,2} \cdots q_{p,m} \tag{5.22}$$

and let

$$I(I_1, \ldots, I_p) = (b_1, b_2 - 1, b_3 - 2, \ldots, b_{mp} - mp + 1)\,.$$

This given, we set

$$SS_{I_1,\ldots,I_p} = (-1)^{\binom{m}{2}\binom{p}{2} + inv(I_1,\ldots,I_p)} S_{I(I_1,\ldots,I_p)}\,. \tag{5.23}$$

Going back to our particular example, we can easily see that to calculate the number of inversions of the permutation

$$0, 6, 4, 7, 2, 11$$

we need only count, for each circle in the diagram of (5.21), the number of circles that are northeast of it, and add all these counts. This gives

$$inv(\Box, \genfrac{}{}{0pt}{}{\Box}{\Box}, \Box\Box) = 0 + 0 + 1 + 2 + 1 + 0 = 4$$

at the same time, we have

$$\binom{m}{2}\binom{p}{2} = \binom{2}{2}\binom{3}{2} = 3$$

and

$$I(I_1, \ldots, I_p) = (0-0,\ 2-1,\ 4-2,\ 6-3,\ 7-4,\ 11-5) = (0,1,2,3,3,6)\,.$$

So we finally obtain in this case

$$SS_{\Box, \genfrac{}{}{0pt}{}{\Box}{\Box}, \Box\Box} = -S_{12336}(x)\,.$$

It is worthwhile noting at this point that, in view of (5.18), the coefficient $c_{I_1, \ldots, I_p}$ does not depend on the order in which $I_1, I_2, \ldots, I_p$ are given. This means that in our particular example we can take advantage of the result in (5.19) and obtain 5 additional terms in the expansion of $S_{113}(x^3)$ by carrying out the above process for each of the remaining permutations of the triplet $(\Box, \genfrac{}{}{0pt}{}{\Box}{\Box}, \Box\Box)$. In the table below we give the resulting diagrams and the corresponding partitions and signs.

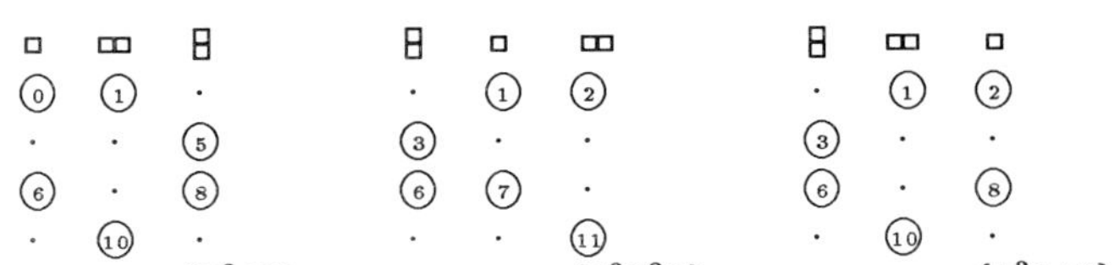

$-$, $\{3^2 45\}$ $\quad$ $+$, $\{1^3 3^2 6\}$ $\quad$ $-$, $\{1^3 345\}$

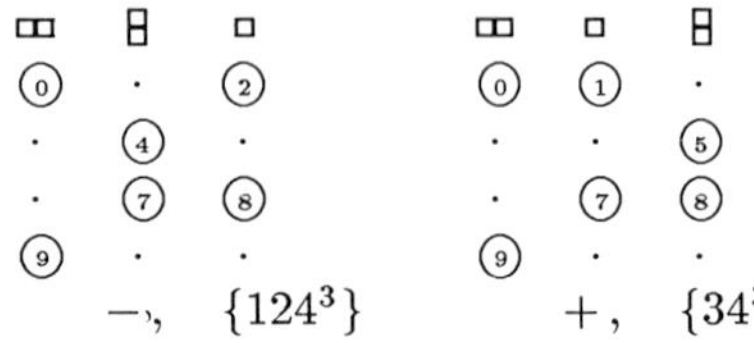

$-$, $\{124^3\}$ $+$, $\{34^3\}$.

We thus obtain that the contribution to the expansion of $S_{113}(x^3)$ coming from the triplet ($\square$, $\boxminus$, $\square\square$) is the expression

$$-2(S_{123^26}(x)+S_{3^245}(x)-S_{1^33^26}(x)+S_{1^345}(x)-S_{124^3}(x)+S_{34^3}(x))\,.$$

Taking all of this into account, we can easily see that the SXP algorithm decomposes into successive applications of the following 3 basic steps. Namely, to calculate the expansion of $S_\mu(x^p)$ when μ is a partition of n, we proceed as follows.

Step 1. We pick a p-tuple of partitions $I_1, I_2, \ldots, I_p$ satisfying the 3 conditions

(a) The Ferrers' diagram of each I_s is contained in that of μ.

(b) $|I_1| \leq |I_1| \leq \cdots \leq |I_p|$

(c) $|I_1| + |I_2| + \cdots + |I_p| = n$.

This done we calculate the coefficient $c_{I_1,\ldots,I_p}$ by the Schur function multiplication algorithm. If this coefficient is not zero we proceed to the next step. Otherwise we repeat step 1.

Step 2. Pick a permutation $I_{\sigma_1}, I_{\sigma_2}, \ldots, I_{\sigma_p}$ of $I_1, I_2, \ldots, I_p$ and construct the labelled circle diagram whose s^{th} column is indexed by I_{σ_1}.

Step 3. Calculate the partition $I(I_{\sigma_1}, I_{\sigma_2}, \ldots, I_{\sigma_p})$ and the corresponding sign.

Step 2. and 3. are to be repeated over all distinct permutations of $I_1, I_2, \ldots, I_p$. This done we go back to step 1 and repeat the process over all possible choices of $I_1, I_2, \ldots, I_p$ satisfying (a) (b) and (c).

In the calculation of the plethysms $S_m[S_n]$ or, more generally, whenever the inner polynomial is $S_n(x)$, we are led to calculate plethysms of the form $S_n[x^p]$. For these, the above algorithm simplifies considerably. We give a separate description of this important special case.

A rather colorful combinatorial algorithm for expanding $S_n(x^p)$ has been given by Chen in [**2**]. To present Chen's algorithm we need further notation and we introduce it with an example. In the figure below we have depicted a Ferrers' diagram F with an attached set of linked circles. We represent this configuration by the symbol $F\,S\ell_{10}(4)$.

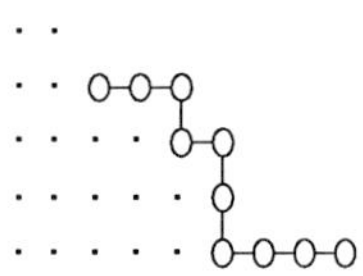

The symbol $S\ell_{10}(4)$ refers to the configuration of linked circles with 10 giving the total number of circles in it and 4 the height of the first circle. We call $S\ell_{10}(4)$ a 10-rim hook starting at 4.

Given a Ferrers' diagram F and an $S\ell_p(i)$ the combined figure $F\,S\ell_p(i)$ is constructed by finding the leftmost position adjacent to the boundary of F at height i then placing the first circle of the rim hook in that position and the remaining $p-1$ circles following in a continuous band *down the staircase*. For instance if F is the Ferrers' diagram corresponding to the partition $I = (1, 3, 5, 6, 9)$ then

$$F\, S\ell_9(5) \;=$$ 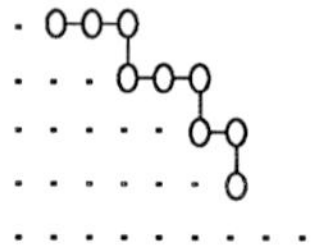

Clearly, F, p and i have to satisfy certain restrictions for the resulting figure to be a Ferrers' diagram. However, in the situations we shall deal with these restrictions will be automatically satisfied.

We shall view the expression $F\, S\ell_p(i)$ as a sort of multiplication, and our ultimate goal is to multiply several of these rim hooks into a *signed* Ferrers' diagram. To this end we define the *sign* of a rim hook H to be $(-1)^{r(H)-1}$ as in section 3. A product of rim hooks is then given a sign which is the product of the signs of its factors. Finally, the resulting configuration will be used to represent a signed Schur function. An example should give the idea of what we have in mind here.

For instance, since

$$S\ell_5(4)S\ell_5(4)S\ell_5(3)S\ell_5(2) \;=$$ 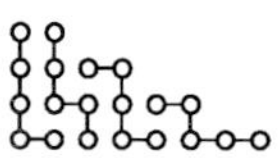

we shall use the symbol $S\ell_5(4)S\ell_5(4)S\ell_5(3)S\ell_5(2)$ to represent $-S_{[2,4,6,8]}$.

This given Chen's algorithm may be summarized by the following formula:

$$S_m(x^p) = \sum_{p\geq i_1\geq i_2\geq\cdots\geq i_m\geq 1} S\ell_p(i_1)S\ell_p(i_2)\cdots S\ell_p(i_m)\,. \tag{5.24}$$

Let us consider the computation of $S_2(x^3)$. Now for $m=2$ and $p=3$ formula (5.24) reduces to

$$\begin{aligned} S_2(x^3) = {} & S\ell_3(3)S\ell_3(3) + S\ell_3(3)S\ell_3(2) + S\ell_3(3)S\ell_3(1) \\ & + S\ell_3(2)S\ell_3(2) + S\ell_3(2)S\ell_3(1) + S\ell_3(1)S\ell_3(1)\,. \end{aligned}$$

Replacing these expressions by the corresponding signed Ferrers' diagrams we get

$S_2(x^3) \;=$

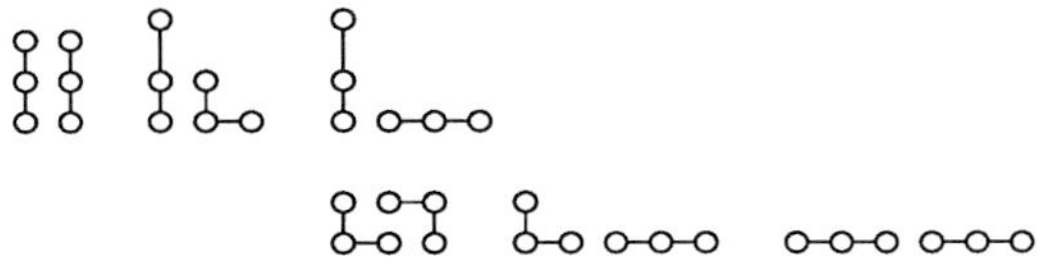

and this is to be interpreted to mean that

$$S_2(x^3) = S_{[2,2,2]} - S_{[1,2,3]} + S_{[1,1,4]} + S_{[3,3]} - S_{[1,5]} + S_{[6]}\,.$$

The combination of the Schur function multiplication rule, Chen's algorithm, plus the recursion (5.16) leads to a reasonable efficient algorithm for computing $S_n[S_m]$. For example, Egecioglu and Remmel [**8**] used this algorithm to produce tables of the expansions of $S_m[S_n]$ which went well beyond those which had appeared previously in the literature. We should also note $S_\lambda[S_n]$ has been computed for $\lambda \vdash 2$ by Littlewood, for $\lambda \vdash 3$ by Thrall [**33**] and Chen, Garsia, and Remmel [**3**], and for $\lambda \vdash 4$ by Foulkes [**11**]. We should also mention a paper by R. Howe [**14**] where a more algebraic method to compute $S_n[S_m]$ is given. Finally, we should mention that Wybourne and his colleagues at the University of Cantebury in Christchurch, New Zealand have developed a nice interactive package called *Schur* which runs on PC's and which computes all the above products of Schur functions plus a great deal more branching rules, etc. for Lie groups.

Bibliography

1. Butler, P. H., and King, R. C., "Branching rules for $U(N) \subseteq U(M)$ and the evaluation of outer plethysms", *J. Math. Phys.* **14** (1973), 741-745.

2. Chen. Y.M., "Combinatorial Algorithms for Plethysm", Doctoral thesis, UCSD (1982).

3. Chen, Y. M., Garsia, A., and Remmel, J. B., "Algorithms for plethysm", *Contemporary Math.*, **34** (1984), 109-153.

4. Duncan, D. G., "Note on a formula by Todd", *J. London Math. Soc.* **27** (1952a), 235-236.

5. Duncan, D. G., "On D. E. Littlewood's algebra of S-functions", *Can. J. Math.* **4** (1952b), 504-512.

6. Duncan, D. G., "Note on the algebra of S-functions", *Can. J. Math.* **6** (1954), 509-510.

7. Egecioglu, O., "Computation of outer products of Schur functions", *Computer Physics Comm.* **28** (1982), 183-187.

8. Egecioglu, O., and Remmel, J., "Symmetric and Antisymmetric Plethysms", *Atomic Data and Nuclear Data Tables*, **32** (1985), 157-196.

9. Foulkes, H. O., "Concomitants of the quintic and sextic up to degree four in coefficients of the ground form", *J. London Math. Soc.* **25** (1950), 205-209.

10. Foulkes, H. O., "The new multiplication of S-functions", *J. London Math. Soc.* **26** (1951), 132-139.

11. Foulkes, H. O., "Plethysm of S-functions", *Philos. Trans. Roy. Soc. London Ser. A* **246** (1954), 555-591.

12. Garsia, A. M. and Remmel, J. B., "Shuffles of permutations and the Kronecker product", *Graphs and Combinatorics* **1** (1985), 217-263.

13. Garsia, A. M. and Remmel, J. B., "Symmetric functions and raising operators", *Linear and Multilinear Algebra* **10** (1981), 15-43.

14. Howe, R., $((GL)_n, (GL)_m)$-duality and Symmetric Plethysm, preprint.

15. James, G. D., The representation theory of the symmetric groups, *Lecture Notes in Math.* **682**, Springer-Verlag, New York, York, 1978.

16. James, G. D., and Kerber, A., "The Representation Theory of the Symmetric Group", Encyc. of Math. and its Appl., Add. Wes. 1981.

17. Littlewood, D. E., "The Theory of Group Characters and Matrix Representations of Groups", Oxford University Press, 1940.

18. Littlewood, D. E., "Invariant theory, tensors and group characters", *Philos. Trans. Roy. Soc. London Ser. A*, **239** (1944), 305-365.

19. Littlewood, D. E., "Modular representations of symmetric groups", *Proc. Roy Soc. London Ser. A*, **209** (1951), 333-353.

20. Littlewood, D. E., "The characters and representations of imprimitive groups", *Proc. London Math. Soc. (3)* **6** (1956), 251-266.

21. Littlewood, D. E., "The Kronecker product of symmetric group representations", *J. London Math. Soc., (1)* **31** (1956), 89-93.

22. Macdonald, I. G., "Symmetric Functions and Hall Polynomials", Oxford Math. Monographs 1979.

23. Murnaghan, F. D., "The analysis of representations of the linear group", *Anais Acad. Brasil. Ci.* **23** (1951), 1-19.

24. Murnaghan, F. D., "On the representations of the symmetric group", *Amer. Journal of Math.*, **59** (1937), 437-488.

25. Murnaghan, F. D., "The characters of the symmetric group", *Anais Acad. Brasil. Ci.* **23** (1951), 141-154.

26. Murnaghan, F. D., "A generalization of Hermite's law of reciprocity", *Anais Acad. Brasil Ci.* **23** (1951), 347-368.

27. Remmel, J. B., "A formula for the Kronecker products of Schur functions of hook shapes", *J. Algebra* **120** (1989), 100-118.

28. Remmel, J. B., "Computing cocharacters of sign trace identities", *Linear and Multilinear Algebra* **23** (1988), 1-14.

29. Remmel, J. B., "On Kronecker products of Schur functions of two row shapes", in preparation.

30. Remmel, J. B. and Whitney, R., "Multiplying Schur functions", *J. of Algorithms* **5** (1984), 471-487.

31. Robinson, G. de B., "On the representations of the symmetric group III", *Amer. J. Math.* **70** (1948), 277-294.

32. Robinson, G. de B., "Induced representations and invariants", *Canad. J. Math.* **2** (1950), 334-343.

33. Thrall, R. M., "On symmetrized Kronecker powers and the structure of the free Lie ring", *Amer. J. Math.* **64** (1942a), 371-388.

34. Thrall, R. M. and Robinson, G. de B., "Supplement to a paper by G. de B. Robinson", *Amer. J. Math.* **73** (1951), 721-724.

35. Todd, J. A., "A note on the algebra of S-functions", *Proc. Cambridge Philos. Soc.* **45** (1949), 328-334.

36. Todd, J. A., "Note on a paper by Robinson", *Canad. J. Math.* **2** (1950), 331-333.

37. Wybourne, B. G., Symmetry Principles in Atomic Spectroscopy, Wiley, New York, 1970.

Acta Applicandae Mathematicae **21**: 137–161, 1990.

POLYNOMIAL IDENTITIES FOR 2 x 2 MATRICES*)

Vesselin Drensky

Institute of Mathematics, Bulgarian Academy of Sciences, Sofia, Bulgaria

0. Introduction

Let $K\langle X\rangle$ be the free associative algebra freely generated by a countable set of symbols $X = \{x_1, x_2, \ldots\}$ over a field K of characteristic 0. The element $f(x_1, \ldots, x_m) \in K\langle X\rangle$ is called a **polynomial identity** for a K-algebra R if $f(r_1, \ldots, r_m) = 0$ for all $r_1, \ldots, r_m \in R$. The set

$$\mathrm{T}(R) = \{f(x_1, \ldots, x_m) \in K\langle X\rangle \mid f \text{ is a polynomial identity for } R\}$$

is a two-sided ideal of $K\langle X\rangle$ which is invariant under all endomorphisms of $K\langle X\rangle$ and is called a **T-ideal.** When $\mathrm{T}(R) \neq 0$, i.e. R satisfies a non-trivial polynomial identity, R is a **PI-algebra.** The class var R of all algebras satisfying the identities from $\mathrm{T}(R)$ is called the **variety of algebras** generated by R.

In the modern treatment of PI-algebras a basic problem is to describe quantitatively the polynomial identities of an algebra R. The natural measures for $\mathrm{T}(R)$ are the multilinear codimension sequence $c_n(R)$, $n = 1, 2, \ldots,$ and the Hilbert (or Poincaré) series $\mathrm{H}(F_m(\mathrm{var}\ R), t_1, \ldots, t_m)$ of the **relatively free algebra** of rank m (or **universal PI-algebra** [20])

$$F_m(\mathrm{var}\ R) = K\langle x_1, \ldots, x_m\rangle / (K\langle x_1, \ldots, x_m\rangle \cap \mathrm{T}(R))$$

in the variety var R. More precisely $c_n(R)$ is the degree of

*) One of the talks given by the author at the Seminar on Computational Algebra of the Department of Mathematics of the II University of Rome (Tor Vergata) in April - May 1990.

AMS subject classifications (1980). 16A38, 20B30, 20G05.

Key words. Polynomial identities, matrices.

the S_n-character $\chi_n(R)$ canonically related with $\mathrm{T}(R)$ and $\mathrm{H}(F_m(\mathrm{var}\ R), t_1, \ldots, t_m)$ is the character of the GL_m-module $F_m(\mathrm{var}\ R)$. Many results, both old and very recent, give explicit values or estimates for $c_n(R)$, $\chi_n(R)$ and $\mathrm{H}(F_m(\mathrm{var}\ R), t_1, \ldots, t_m)$.

Especially attractive is the quantitative theory of the polynomial identities of the algebra M_k of the $k \times k$ matrices. Since the algebra $F_m(\mathrm{var}\ M_k)$ is isomorphic to the algebra generated by m generic $k \times k$ matrices, in the study of $\mathrm{T}(M_k)$ the invariant theory of matrices [14, 17] was successfully involved. We refer to [8] for a survey on the results before 1982.

Up till now the results on $\mathrm{T}(M_k)$, $k > 2$, are far from their final form and only the structure of the identities of the 2×2 matrices is well known. In a series of papers (see [19] for a survey) Regev obtained the asymptotic behaviour of $c_n(M_2)$ and $\chi_n(M_2)$; the explicit expression of $\chi_n(M_2)$ was found by Formanek [9] and the author [5]; Procesi [15] computed the codimension sequence of M_2. Formanek, Halpin and Li [10] for $m = 2$ and Procesi [15], Formanek [9] and the author [5] in the general case computed the Hilbert series of $F_m(\mathrm{var}\ M_2)$.

One of the possible ways to study $\mathrm{T}(M_2)$ is via the generic trace algebra generated by m generic 2×2 matrices together with all the traces of the generic matrix algebra $F_m(\mathrm{var}\ M_2)$ and the approach of Regev, Formanek and Procesi followed this way. For a survey on the 2×2 generic trace algebra see [12].

The purpose of the present paper is to give a revised version of the quantitative results on $T(M_2)$ without any invariant theory. The proofs have been scattered in a series of papers of the author [2, 5, 6]. The methods developed work also for other PI-algebras and we refer to the survey [7] for further applications.

1. Preliminaries

All algebras, vector spaces and tensor products are over a fixed field K of characteristic 0. We consider unitary algebras only. For a background of the theory of PI-algebras we refer to [20]. Let P_n be the set of all multilinear elements of degree n in $K\langle x_1,\dots,x_n\rangle \subset K\langle X\rangle$. It is well known that the polynomial identities for an algebra R are determined by the multilinear ones, i.e. as a T-ideal $T(R)$ is generated by $P_n \cap T(R)$, $n = 1,2,\dots$. It turns out that $\dim (P_n \cap T(R)) \approx \dim P_n = n!$ and it is more reasonable to study the n-th (multilinear) **codimension** of $T(R)$

$$c_n(T(R)) = c_n(R) = \dim P_n/(P_n \cap T(R)),\ n = 1,2,\dots .$$

By a theorem of Regev [18] there exists a real constant a such that $c_n(R) \le a^n$.

Let S_n be the symmetric group of degree n acting on the set of symbols $\{1,\dots,n\}$. The vector space P_n is a left S_n-module with an action of S_n defined by

$$\sigma(x_{i_1}\dots x_{i_n}) = x_{\sigma(i_1)}\dots x_{\sigma(i_n)},$$

where $x_{i_1}\dots x_{i_n} \in P_n$ and $\sigma \in S_n$. It is easy to see that $P_n \cap T(R)$ is an S_n-submodule of P_n and the S_n-character

$$\chi_n(R) = \chi(P_n/(P_n \cap T(R)))$$

is called the n-th **cocharacter** of $T(R)$ (or of R).

It is known that the irreducible characters of every finite group G (over a sufficiently large field) are in a one-to-one correspondence with the conjugacy classes of G. For $G = S_n$ the conjugacy class σ^{S_n}, $\sigma \in S_n$, is determined by the cycle type of σ. Assume that σ is presented as a product of independent cycles

$$\sigma = \underbrace{(*\ldots*)}_{\lambda_1}\ldots\underbrace{(*\ldots*)}_{\lambda_r}, \quad \lambda_1 \geq \ldots \geq \lambda_r > 0, \quad \lambda_1 + \ldots + \lambda_r = n.$$

Then we relate a **partition** $\lambda = (\lambda_1,\ldots,\lambda_r)$ of n (notation $\lambda \vdash n$) and a **Young diagram** $[\lambda]$. For example,

$$[\lambda] = [4,3^2,2,1] =$$ 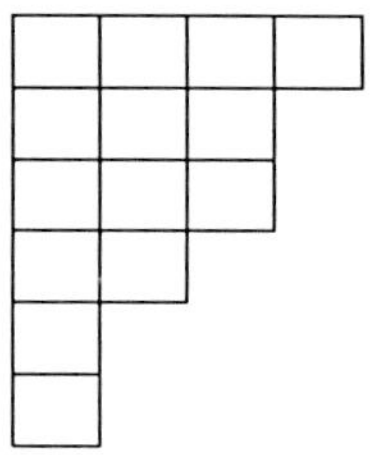 .

We denote by $\chi(\lambda)$ and $M(\lambda)$, respectively, the irreducible S_n-character and module related with the partition λ. Then

$$\chi_n(R) = \sum \kappa(\lambda)\chi(\lambda)$$

and one of the main problems for the polynomial identities of R is to describe the **multiplicities** $\kappa(\lambda)$.

The degree of the irreducible S_n-characters are given by the **hook formula**. Let $\lambda = (\lambda_1,\ldots,\lambda_r)$ be a partition of n and let $\lambda'_1,\ldots,\lambda'_s$ be the lengths of the columns of the diagram $[\lambda]$.

Then

$$\deg \chi(\lambda) = \dim M(\lambda) = \frac{n!}{\Pi(\lambda_i+\lambda'_j-i-j+1)}.$$

The vector space V_m with a basis $x_1,\ldots,x_m$ is a left GL_m-module with the canonical action of the general linear group GL_m. The free algebra $K<x_1,\ldots,x_m>$ is isomorphic to the tensor algebra of V_m. Therefore, $K<x_1,\ldots,x_m>$ is a GL_m-module with the diagonal action of GL_m defined by

$$g(x_{i_1}\ldots x_{i_n}) = g(x_{i_1})\ldots g(x_{i_n}),\ x_{i_1}\ldots x_{i_n} \in A_m^{(n)},\ g \in GL_m.$$

It is well known (we refer to [11] and [23] for details) that the irreducible polynomial representations of GL_m are described by partitions and are equivalent to subrepresentations of GL_m in $K<x_1,\ldots,x_m>$. For $\lambda = (\lambda_1,\ldots,\lambda_r) \vdash n$, $r \le m$, there exists an irreducible GL_m-submodule $N_m(\lambda)$ of $K<x_1,\ldots,x_m>$ and, when $m \ge n$, the multilinear component $P_n \cap N_m(\lambda)$ is isomorphic to the S_n-submodule $M(\lambda)$ of P_n. As a consequence, for every PI-algebra R the vector space $K<x_1,\ldots,x_m> \cap \mathrm{T}(R)$ is a GL_m-submodule of $K<x_1,\ldots,x_m>$ and the structures of the S_n-modules $P_n \cap \mathrm{T}(R)$, $n = 1,2,\ldots,$ and the GL_m-module $K<x_1,\ldots,x_m> \cap \mathrm{T}(R)$ are the same [1, 2]. More precisely, let

$$F_m(\mathrm{var}\ R) = K<x_1,\ldots,x_m>/(K<x_1,\ldots,x_m> \cap \mathrm{T}(R))$$

be the relatively free algebra of rank m in var R and let $P_n/(P_n \cap \mathrm{T}(R)) = \sum_{\lambda\vdash n} \kappa(\lambda)M(\lambda)$, $n = 1,2,\ldots$. Then

$$F_m(\mathrm{var}\ R) = \sum_n \sum_{\lambda\vdash n} \kappa(\lambda)N_m(\lambda).$$

In particular, the multiplicity of $N_m(\lambda)$ in $F_m(\mathrm{var}\ R)$ does not depend on the rank m. The role of the character of a polynomial representation of GL_m is played by its Hilbert series. The vector space $K\langle x_1,\ldots,x_m\rangle$ has a grading counting the degree in any variable x_i in the monomials from $K\langle x_1,\ldots,x_m\rangle$ and this grading is inherited by $F_m(\text{var } R)$. Let $F_m^{(n_1,\ldots,n_m)}(\text{var } R)$ be the multihomogeneous component of degree n_i in x_i, $i = 1,\ldots,m$. The formal power series

$$H(F_m(\text{var } R), t_1,\ldots,t_m) = \sum \dim F_m^{(n_1,\ldots,n_m)}(\text{var } R)\, t_1^{n_1}\ldots t_m^{n_m}$$

is the **Hilbert** (or Poincaré) **series** of $F_m(\text{var } R)$. The character associated with the irreducible module $N_m(\lambda)$ is a symmetric polynomial and is called the **Schur function** $S_\lambda(t_1,\ldots,t_m)$ corresponding to the partition λ (see [13, Ch.1]). For example, in the two extreme cases $\lambda = (n)$ and $\lambda = (1^n)$ the Schur functions are the n-th complete symmetric function h_n and the n-th elementary symmetric function e_n, respectively; when $\lambda = (m^k)$, then $S_\lambda(t_1,\ldots,t_m) = (t_1\ldots t_m)^k$. The GL_m-module $F_m(\text{var } R)$ is uniquely determined by its Hilbert series and if

$$H(F_m(\text{var } R), t_1,\ldots,t_m) = \sum_n \sum_{\lambda\vdash n} \kappa(\lambda) S_\lambda(t_1,\ldots,t_m),$$

we derive for $n \le m$

$$P_n/(P_n \cap T(R)) = \sum_{\lambda\vdash n} \kappa(\lambda) M(\lambda).$$

The tensor product of GL_m-modules is also a GL_m-module. The Littlewood-Richardson rule gives an algorithm for computing the multiplicities of the irreducible components of the tensor product; for details we refer to [13, Ch.1]. In particular, we make use of the following special case of this rule.

Proposition 1.1.

$$N_m(\lambda_1,\ldots,\lambda_m) \otimes N_m(q) \cong \sum N_m(\lambda_1 + p_1,\ldots,\lambda_m + p_m),$$

where the summation is over all non-negative integers $p_1,\ldots,p_m$ *such that* $p_1 + \ldots + p_m = q$ *and* $\lambda_i + p_i \le \lambda_{i-1}$, $i = 2,\ldots,m$.■

When $q = 1$ this is the well known Branching Theorem. For example, below we give the decomposition when λ is a partition in three parts.

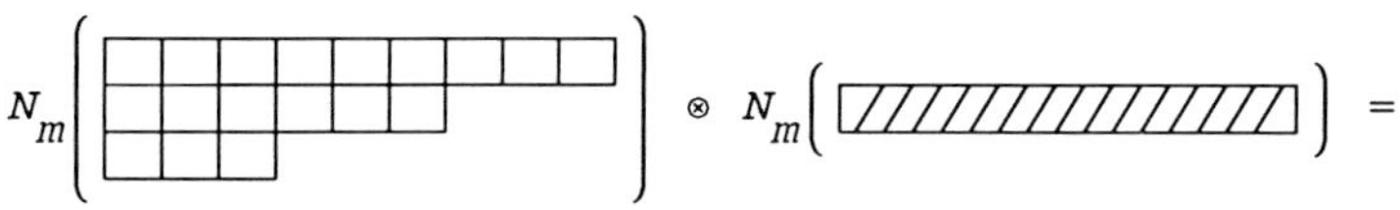

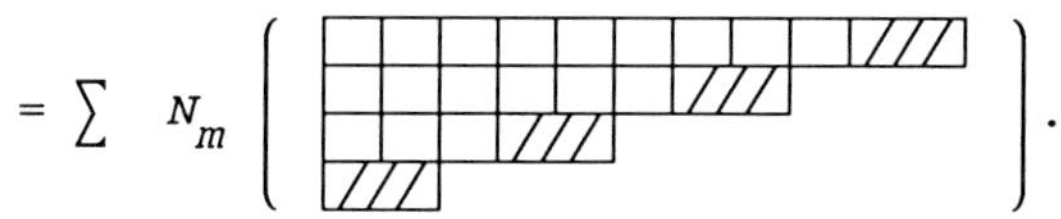

Here the summation is over all possible diagrams containing the left diagram such that every two shaded boxes belong to different columns.

We also use the symmetrized tensor power $N^{\otimes^s k}$ of a GL_m-module N identifying the tensors $v_1 \otimes \ldots \otimes v_k$ and $v_{\sigma(1)} \otimes \ldots \otimes v_{\sigma(k)}$, $\sigma \in S_k$. The relation between the GL_m-characters of N and $N^{\otimes^s k}$ is a very special case of the general notion of plethysm $f \odot g$ of two symmetric functions $f \odot g$ (see [11, 13]). A result of Thrall [22] (see also [13, Exercise 5, p.45]) gives that

$$N_m(1^2)^{\otimes^s k} = \sum N_m(\lambda_1^2,\ldots,\lambda_r^2), \quad \lambda = (\lambda_1,\ldots,\lambda_r) \vdash k.$$

In the language of plethysm this is interpreted as

$$\sum_k \sum_\lambda S_{(\lambda_1^2,\ldots,\lambda_r^2)} = \sum_k h_k \odot e_2 = \prod_{i<j} (1 - t_i t_j)^{-1}.$$

We define a right action of S_n on the vector space $K\langle x_1,\ldots,x_m\rangle^{(n)}$ of all homogeneous polynomials of degree n from $K\langle x_1,\ldots,x_m\rangle$ by

$$(x_{i_1}\ldots x_{i_n})\sigma^{-1} = x_{i_{\sigma(1)}}\ldots x_{i_{\sigma(n)}}, \quad \sigma \in S_n,$$

i.e. S_n acts on the positions of the variables in the monomials of degree n. If $\lambda = (\lambda_1,\ldots,\lambda_r) \vdash n$, $\lambda_r \neq 0$, and $N_m(\lambda) \subset K\langle x_1,\ldots,x_m\rangle^{(n)}$, then $N_m(\lambda)$ is generated by a non-zero polynomial

$$f = f_\lambda = \left(\prod_{i=1}^{\lambda_1} s_{r_i}(x_1,\ldots,x_{r_i})\right)\sum a_\sigma \sigma = \sum a_\sigma f_\sigma, \quad a_\sigma \in K,\ \sigma \in S_n.$$

Here $r_1 = r, r_2,\ldots,r_{\lambda_1}$ are the lengths of the columns of the diagram $[\lambda]$,

$$s_p(x_1,\ldots,x_p) = \sum (\text{sign } \tau) x_{\tau(1)}\ldots x_{\tau(p)}, \quad \tau \in S_p$$

is the **standard polynomial** and

$$f_\sigma = \prod_{i=1}^{\lambda_1} s_{r_i}(x_1,\ldots,x_{r_i})\sigma.$$

The polynomial f_λ is called a **standard generator** of the module $N_m(\lambda)$. Its complete linearization generates an irreducible S_n-module $M(\lambda)$ and every submodule $M(\lambda) \subset P_n$ can be obtained in this way [23]. Therefore, f_λ is a standard generator if $\deg_{x_i} f_\lambda = \lambda_i$, $i = 1,\ldots,r$, and f_λ is a sum of polynomials f_σ, where every variable x_i participates in a skew-symmetric sum with the variables $x_1,\ldots,x_{i-1}$. For example,

$$f_{(2,1)} = a_{(12)}(x_1x_2 - x_2x_1)x_1 + a_{(23)}x_1(x_1x_2 - x_2x_1),$$

where $a_{(12)}, a_{(23)} \in K$, $(a_{(12)},a_{(23)}) \neq (0,0)$.

2. Unitary Algebras

We define (higher) commutators by

$$[u_1,u_2] = u_1(\mathrm{ad}u_2) = u_1u_2 - u_2u_1,$$

$$[u_1,\ldots,u_{n-1},u_n] = [u_1,\ldots,u_{n-1}](\mathrm{ad}u_n),\ n > 2.$$

Every associative algebra R has a structure of a Lie algebra with the new multiplication $[u_1,u_2]$, $u_1,u_2 \in R$. The Lie subalgebra L_m of $K\langle x_1,\ldots,x_m\rangle$ generated by the free generators $x_1,\ldots,x_m$ is isomorphic to the free Lie algebra of rank m. Let $x_1,\ldots, x_m, u_1, u_2,\ldots$ be an ordered basis of L_m, where $u_1,u_2,\ldots$ are higher commutators. Since $K\langle x_1,\ldots,x_m\rangle$ is the universal enveloping algebra of L_m, the Poincaré-Birkhoff-Witt theorem gives that $K\langle x_1,\ldots,x_m\rangle$ has a basis

$$\{x_1^{a_1}\ldots x_m^{a_m} u_1^{b_1}\ldots u_n^{b_n} \mid a_i, b_j \geq 0\}.$$

Lemma 2.1. *Let*

$$f(x_1,\ldots,x_m) = \sum_a \sum_b p_{ab}x_1^{a_1}\ldots x_m^{a_m} u_1^{b_1}\ldots u_n^{b_n} \in K\langle x_1,\ldots,x_m\rangle$$

be a polynomial identity for an algebra R, *where* $a = (a_1,\ldots,a_m)$, $b = (b_1,\ldots,b_n)$, $p_{ab} \in K$ *and* $u_1,\ldots,u_n$ *are higher commutators from the basis of* L_m. *Then for every fixed* $a = (a_1,\ldots,a_m)$,

$$f_a(x_1,\ldots,x_m) = \sum_b p_{ab}u_1^{b_1}\ldots u_n^{b_n}$$

is also a polynomial identity for R.

Proof. Since $u_j = u_j(x_1,\ldots,x_m)$ are higher commutators,

$$u_j(x_1 + q, x_2,\ldots,x_m) = u_j(x_1,x_2,\ldots,x_m),\quad q \in K.$$

Hence

$$f(x_1 + q, x_2,\ldots,x_m) = \sum_a\sum_b p_{ab}(x_1 + q)^{a_1}x_2^{a_2}\ldots x_m^{a_m}u_1^{b_1}\ldots u_n^{b_n} =$$

$$= \sum_a\sum_b\sum_k p_{ab}\binom{a_1}{k}q^k x_1^{a_1-k}x_2^{a_2}\ldots x_m^{a_m}u_1^{b_1}\ldots u_n^{b_n}.$$

Let $a_1^{(0)}$ be the maximal integer with $p_{ab} \neq 0$. Since the field K is infinite, standard Vandermonde arguments [20, Corollary 2.3.31, p.131] show that

$$\sum_{a'}\sum_b p_{a'b}x_2^{a_2}\ldots x_m^{a_m}u_1^{b_1}\ldots u_n^{b_n},\quad a' = (a_1^{(0)},a_2,\ldots,a_m),$$

is also a polynomial identity for R. Now the proof is completed by an easy induction on a_1 and then on the other a_i, $i = 2,\ldots,m$.■

We denote by B_m the vector subspace of $K\langle x_1,\ldots,x_m\rangle$ spanned by all products of higher commutators $u_1^{b_1}\ldots u_n^{b_n}$, $b_i \geq 0$. The elements of B_m are called **proper** (or **commutator**) **polynomials.** It follows from the proof of Lemma 2.1 that $f(x_1,\ldots,x_m) \in K\langle x_1,\ldots,x_m\rangle$ belongs to B_m if and only if the formal derivatives $\partial f/\partial x_j$ equal zero, $j = 1,\ldots,m$. Let $\Gamma_n = B_n \cap P_n$ be the set of all multilinear proper polynomials. A basis of Γ_n was given in [21]; it can be obtained applying the argument that the commutators $[x_n,x_{\sigma(1)},\ldots,x_{\sigma(n-1)}]$, $\sigma \in S_{n-1}$, form a basis of $P_n \cap L_n$, $n \geq 2$.

For every algebra R the T-ideal $T(R)$ is generated by its multilinear elements and Lemma 2.1 gives that all polynomial identities of R are consequences of $\Gamma_n \cap T(R)$, $n = 2,3,\ldots$, [21]. Clearly, the GL_m-module $B_m \cap T(R)$ and the S_n-modules $\Gamma_n \cap T(R)$, $n \leq m$, have a similar structure and the same holds for $B_m/(B_m \cap T(R))$ and $\Gamma_n/(\Gamma_n \cap T(R))$. For many algebras R the description of the GL_m-module $B_m/(B_m \cap T(R))$ is simpler than the description of $F_m(\text{var } R)$. Our strategy is to study $B_m/(B_m \cap T(R))$ and as a consequence to derive quantitative results on $T(R)$.

Theorem 2.2. [5] *Let R be a PI-algebra.*

(i) *If $\{f_1, f_2,\ldots\}$ is a basis of $B_m/(B_m \cap T(R))$, then $F_m(\text{var } R)$ has a basis*

$$\{x_1^{a_1}\ldots x_m^{a_m} f_j \mid a_i \geq 0,\ j = 1,2,\ldots\}.$$

(ii) *The Hilbert series of $B_m/(B_m \cap T(R))$ and $F_m(\text{var } R)$ satisfy the relation*

$$H(F_m(\text{var } R), t_1,\ldots,t_m) = \prod_{i=1}^{m}(1 - t_i)^{-1} H(B_m/(B_m \cap T(R)), t_1,\ldots,t_m).$$

(iii) *The following GL_m-module isomorphism holds*

$$F_m(\text{var } R) \cong K[x_1,\ldots,x_m] \otimes (B_m/(B_m \cap T(R)),$$

where $K[x_1,\ldots,x_m]$ is the algebra of the polynomials in commuting variables with the canonical GL_m-action.

(iv) *If $B_m/(B_m \cap T(R)) = \sum \kappa_1(\mu) N_m(\mu)$ and $F_m(\text{var } R) = \sum \kappa(\lambda) N_m(\lambda)$, then $\kappa(\lambda) = \sum \kappa_1(\mu)$, where for $\lambda = (\lambda_1,\ldots,\lambda_m)$ the summation runs over all partitions*

$\mu = (\mu_1,\ldots,\mu_m)$ *such that* $\lambda_1 \geq \mu_1 \geq \ldots \geq \lambda_m \geq \mu_m$.

Proof. The statement of (i) follows immediately from Lemma 2.1. As a consequence of (i) we derive that the graded vector spaces $F_m(\text{var } R)$ and $K[x_1,\ldots,x_m] \otimes (B_m/(B_m \cap T(R))$ are isomorphic. If N' and N'' are graded vector spaces, then the Hilbert series of $N' \otimes N''$ is a product of the Hilbert series of N' and N''. This gives (ii) because in the neighbourhood of $(0,\ldots,0) \in \mathbb{R}^m$ the Hilbert series of $K[x_1,\ldots,x_m]$ is expressed as

$$H(K[x_1,\ldots,x_m],t_1,\ldots,t_m) = \prod_{i=1}^{m}(1 - t_i)^{-1}.$$

Since the GL_m-modules are determined by their Hilbert series, (iii) is equivalent to (ii). Finally, (iv) follows from (iii) and Proposition 1.1 because $K[x_1,\ldots,x_m] = \Sigma\, N_m(n)$.■

Theorem 2.3. [5] *Let* R *be a PI-algebra.*

(i) *If* $\{g_{k1},\ldots g_{k\gamma_k}\}$ *is a basis of* $\Gamma_k/(\Gamma_k \cap T(R))$, *then* $P_n/(P_n \cap T(R))$ *has a basis consisting of all multilinear polynomials of the form*

$$x_{i_1}\ldots x_{i_{n-k}} g_{kp}(x_{j_1},\ldots,x_{j_k}),$$

where $i_1 < \ldots < i_{n-k}$, $j_1 < \ldots < j_k$ *and* $p = 1,\ldots,\gamma_k$, $k = 0,1,\ldots,n$.

(ii) *The codimension sequence* $c_n(R)$, $n = 0,1,2,\ldots,$ *and the* **proper codimension sequence** $\gamma_k(R) = \dim \Gamma_k/(\Gamma_k \cap T(R))$, $k = 0,1,2,\ldots,$ *are related by the equality*

$$c_n(R) = \sum \binom{n}{k} \gamma_k(R).$$

(iii) *The* **codimension series** $c(R,t) = \sum c_n(R)\,t^n$ *and the*

proper codimension series $\gamma(R,t) = \sum \gamma_k(R)t^k$ *satisfy the equation*

$$c(R,t) = \frac{1}{1-t}\gamma(R, \frac{t}{1-t}).$$

Proof. Since $P_n/(P_n \cap T(R))$ and $\Gamma_k/(\Gamma_k \cap T(R))$ are the multilinear components of $F_m(\text{var } R)$ and $B_m/(B_m \cap T(R))$, respectively, (i) is a consequence of Theorem 2.2 (i). The proof of (ii) follows by counting the elements from the basis of $P_n/(P_n \cap T(R))$ in (i). Finally, let a_n, $n = 0,1,2,\dots$, and b_k, $k = 0,1,2,\dots$, be two sequences related with $a_n = \sum \binom{n}{k} b_k$. It is easy to see that the generating functions $a(t) = \sum a_n t^n$ and $b(t) = \sum b_k t^k$ satisfy $a(t) = \frac{1}{1-t} b(\frac{t}{1-t})$ and this gives (iii).■

3. Proper Polynomial Identities for 2 x 2 Matrices

Let $M_2(K[\zeta])$ be the 2×2 matrix algebra with entries from the polynomial algebra

$$K[\zeta] = K[\zeta_{pq}^{(i)} \mid p, q = 1,2, i = 1,\dots,m]$$

The subalgebra of $M_2(K[\zeta])$ generated by the **generic matrices**

$$y_i = \begin{pmatrix} \zeta_{11}^{(i)} & \zeta_{12}^{(i)} \\ \zeta_{21}^{(i)} & \zeta_{22}^{(i)} \end{pmatrix}, \quad i = 1,\dots,m,$$

is the **generic 2 x 2 matrix algebra.** A polynomial $f(x_1,\dots,x_m)$ vanishes on $y_1,\dots,y_m$ if and only if $f(v_1,\dots,v_m) = 0$ for all matrices $v_1,\dots,v_m$ with entries from K. Therefore the relatively free algebra $F_m(\text{var } M_2)$ is isomorphic to the generic 2×2 matrix algebra. There is a standard diagonalization trick due to Procesi which allows to consider one of the generic matrices diagonal [20, Proposition 1.3.15, p.17]. Here we do one more step. Let

$$z_1 = \begin{pmatrix} \xi_1 & 0 \\ 0 & \xi_2 \end{pmatrix}, \quad z_2 = \begin{pmatrix} \eta_1 & \eta \\ \eta & \eta_2 \end{pmatrix}, \quad z_i = y_i, \quad i = 3,\dots,m,$$

be matrices with entries from the polynomial algebra

$$\Xi = K[\xi_1, \xi_2, \eta_1, \eta_2, \eta, \zeta_{pq}^{(i)} \mid p, q = 1,2, \ i = 3,\dots,m],$$

i.e. the first two matrices z_1 and z_2 are diagonal and symmetric, respectively. The following assertion is a generalization of [20, Proposition 1.3.15, p.17]. We sketch the proof only.

Lemma 3.1. *The subalgebra of* $M_2(K[\Xi])$ *generated by the matrices* $z_1,\dots,z_m$ *is isomorphic to the generic* 2×2 *matrix algebra.*

Proof. Let $\Omega = \overline{K(\zeta)}$ be the algebraic closure of the field of fractions of $K[\zeta]$. Since the generic matrix y_1 has different characteristic values, there exists an invertible matrix u_1 from $M_2(\Omega)$ such that

$$u_1^{-1} y_1 u_1 = \begin{pmatrix} \rho_1 & 0 \\ 0 & \rho_2 \end{pmatrix} \text{ is diagonal and } u_1^{-1} y_2 u_1 = \begin{pmatrix} \sigma_{11} & \sigma_{12} \\ \sigma_{21} & \sigma_{22} \end{pmatrix}.$$

Clearly σ_{12} and σ_{21} are different from 0. Let $u_2 = \tau_1 e_{11} + \tau_2 e_{22} \in M_2(\Omega)$ be an invertible matrix. Then

$$u_2^{-1} u_1^{-1} y u_1 u_2 = u_1^{-1} y u_1 \quad \text{and} \quad u_2^{-1} u_1^{-1} y_2 u_1 u_2 = \begin{pmatrix} \sigma_{11} & \tau_1^{-1}\tau_2\sigma_{12} \\ \tau_2^{-1}\tau_1\sigma_{21} & \sigma_{22} \end{pmatrix}.$$

The field Ω is algebraically closed and we can choose τ_1 and τ_2 such that $\tau_1^{-1}\tau_2\sigma_{12} = \tau_2^{-1}\tau_1\sigma_{21}$. The matrices $u_2^{-1} u_1^{-1} y_i u_1 u_2$ are obtained by specialization of the entries of the matrices z_i, $i = 1,\dots,m$. Therefore, the mapping

$$\begin{array}{ccc} y_i & \longrightarrow & z_i \\ & \searrow \quad \swarrow & \\ & u_2^{-1}u_1^{-1}y_iu_1u_2 & \end{array}$$

gives rise to the commutative diagram

$$\begin{array}{ccc} \langle y_i \mid i = 1,\ldots,m\rangle & \xrightarrow{\ \vartheta\ } & \langle z_i \mid i = 1,\ldots,m\rangle \\ & \varphi \searrow \quad \swarrow \psi & \\ & \langle u_2^{-1}u_1^{-1}y_iu_1u_2 \mid i = 1,\ldots,m\rangle & \end{array}$$

where $\langle\ldots\rangle$ is the algebra generated by the set $\{\ldots\}$. Since ϑ and ψ are epimorphisms and φ is an isomorphism, we derive that ϑ is also an isomorphism.∎

Till the end of the section we modify the exposition of [2, Section 4] (see also [7, Section 3]). Let

$$w_1 = \begin{pmatrix} 1 & 0 \\ 0 & -1 \end{pmatrix}, \quad w_2 = \begin{pmatrix} 0 & 1 \\ 1 & 0 \end{pmatrix}, \quad w_3 = \begin{pmatrix} 0 & 1 \\ -1 & 0 \end{pmatrix}.$$

Theorem 3.2. *Let $f_\mu(x_1,\ldots,x_r)$, $\mu = (\mu_1,\ldots,\mu_r)$, be a standard generator of the submodule $N_m(\mu)$ of the GL_m-module B_m of the proper polynomials in $K\langle x_1,\ldots,x_m\rangle$. If $\mu_4 \neq 0$, then $f_\mu(x_1,\ldots,x_r)$ is a polynomial identity for M_2. When $\mu_4 = 0$, then $f_\mu(x_1,x_2,x_3)$ is a polynomial identity for M_2 if and only if $f_\mu(w_1,w_2,w_3) = 0$, where the matrices w_1,w_2,w_3 are defined above.*

Proof. Every matrix $v \in M_2$ is presented as $v = v_0 + \frac{1}{2}\operatorname{tr} v$, where v_0 is a traceless matrix. Since f_μ is a standard generator, every variable x_i participates in f_μ

in a skew-symmetric sum with the variables $x_1,\dots,x_{i-1}$;

$$f_\mu(x_1,\dots,x_i + a,\dots,x_r) = f_\mu(x_1,\dots,x_i,\dots,x_r)$$

for all $a \in K$ because $f_\mu \in B_m$. Therefore, it suffices to verify that $f_\mu(v_1,\dots,v_r) = 0$ for linearly independent matrices $v_1,\dots,v_r$ from the vector space sl_2 of all traceless 2×2 matrices. Since $\dim sl_2 = 3$, the skew-symmetry of f_μ gives that $f_\mu(x_1,\dots,x_r)$ is a polynomial identity for M_2 when $r > 3$. Now, let $r \le 3$. In virtue of Lemma 3.1, $f_\mu(x_1,x_2,x_3) \in T(M_2)$ if and only if $f_\mu(z_1,z_2,z_3) = 0$ for the generic matrices z_1,z_2,z_3. The matrices $1, w_1, w_2, w_3$ form a basis of M_2. We rewrite z_1,z_2,z_3 as their linear combinations:

$$z_1 = \phi_{01}1 + \xi w_1,\ z_2 = \phi_{02}1 + \phi_{12}w_1 + \eta w_2,$$

$$z_3 = \phi_{30}1 + \phi_{31}w_1 + \phi_{32}w_2 + \zeta w_3,$$

where ϕ_{ij}, ξ, η, ζ are independent commuting variables. The polynomial f_μ is multihomogeneous of degree μ_i in x_i and, by the skew-symmetry of f_μ

$$f_\mu(z_1,z_2,z_3) = f_\mu(\xi w_1,\eta w_2,\zeta w_3) = \xi^{\mu_1}\eta^{\mu_2}\zeta^{\mu_3}f_\mu(w_1,w_2,w_3)$$

and $f_\mu(z_1,z_2,z_3) = 0$ if and only if $f_\mu(w_1,w_2,w_3) = 0$.■

Remark 3.3. *The matrices* w_1,w_2,w_3 *in Theorem* 3.2 *can be replaced by the matrices*

$$h_1 = -\frac{1}{2}(e_{11} - e_{22})\sqrt{-1},\ h_2 = \frac{1}{2}(e_{12} + e_{21})\sqrt{-1},\ h_3 = \frac{1}{2}(e_{12} - e_{21}).$$

The table of multiplication of these matrices is the following

$$h_1^2 = h_2^2 = h_3^2 = -1/4,\ h_1h_2 = -h_2h_1 = h_3/2,$$

$$h_2h_3 = -h_3h_2 = h_1/2,\ h_3h_1 = -h_1h_3 = h_2/2$$

and this simplifies the computations. In particular, it is easy to see that *for* $\mu = (\mu_1, \mu_2, \mu_3)$ *and* $f_\mu \in B_m$ *there exists a constant* $a(f_\mu) \in K$ *such that* $f_\mu(h_1, h_2, h_3) = a(f_\mu) h_1^{\delta_1} h_2^{\delta_2} h_3^{\delta_3}$, *where* $\delta_i = 0,1$ *and* $\delta_i \equiv \mu_i \pmod 2$, $i = 1,2,3$.■

The following theorem describes the GL_m-structure of the proper polynomials in $F_m(\mathrm{var}\, M_2)$.

Theorem 3.4. $B_m/(B_m \cap T(M_2)) = \sum N_m(\mu_1, \mu_2, \mu_3)$, *where the summation is over all partitions* $\mu = (\mu_1, \mu_2, \mu_3)$ *different from* $\mu = (1^3)$ *and* $\mu = (n)$, $n \geq 1$.

Proof. Let $B_m/(B_m \cap T(M_2)) = \sum \kappa_1(\mu) N_m(\mu)$. By Theorem 3.2 $\kappa_1(\mu) = 0$ if $\mu = (\mu_1, \ldots, \mu_r)$ and $\mu_4 \neq 0$. Now, let f'_μ and f''_μ be standard generators of two isomorphic GL_m-submodules $N_m(\mu)$ of $B_m \subset K\langle x_1, \ldots, x_m\rangle$ and let $f'_\mu \notin (B_m \cap T(M_2)$. Then, for the constants $a(f'_\mu)$, $a(f''_\mu) \in K$ from Remark 3.3

$$f_\mu = a(f''_\mu) f'_\mu - a(f'_\mu) f''_\mu \in B_m, \qquad f_\mu(h_1, h_2, h_3) = 0.$$

Therefore, f'_μ and f''_μ are linearly dependent modulo $B_m \cap T(M_2)$ and $\kappa_1(\mu) \leq 1$. The proof will be completed if we show that $\kappa_1(\mu) = 0$ for $\mu = (1^3)$ and $\mu = (n)$, $n \geq 1$ and if we construct non-zero standard generators from $B_m/(B_m \cap T(M_2))$ for all other partitions $\mu = (\mu_1, \mu_2, \mu_3)$. The multiplicity of $N_m(n)$ in $K\langle x_1, \ldots, x_m\rangle$ is 1 and the corresponding standard generator is x_1^n. Clearly $x_1^n \notin B_m$ for $n \geq 1$ and, therefore $\kappa_1(n) = 0$. Similarly, $\kappa_1(1^3) = 0$ because the only submodule $N_m(1^3)$ in $K\langle x_1, \ldots, x_m\rangle$ is generated by the standard polynomial $s_3(x_1, x_2, x_3)$ which does not belong to B_m. Now, let $\mu = (\mu_1, \mu_2, \mu_3)$ be another partition. Since for $\mu = (0)$ $N_m(\mu) = K$, we assume that $\mu \vdash n > 1$. Let

$$f_\mu = s_2(x_1,x_2)(\mathrm{ad}x_1)^{\mu_1-\mu_2}(s_3(\mathrm{ad}x_1,\mathrm{ad}x_2,\mathrm{ad}x_3))^{\mu_3}(s_2(x_1,x_2))^{\mu_2-\mu_3-1}$$

when $\mu_2 \neq \mu_3$ and, for $\mu_2 = \mu_3$,

$$f_\mu = \sum(\mathrm{sign}\ \sigma)x_{\sigma(1)}(\mathrm{ad}x_1)^{\mu_1-\mu_3}(s_3(\mathrm{ad}x_1,\mathrm{ad}x_2,\mathrm{ad}x_3))^{\mu_3-1}[x_{\sigma(2)},x_{\sigma(3)}].$$

The polynomial f_μ has the desired skew-symmetry and all variables are in commutators. Hence f_μ is a standard generator of $N_m(\mu) \subset B_m$. Direct verifications show that $f_\mu(h_1,h_2,h_3) \neq 0$ for the matrices h_1,h_2,h_3 from Remark 3.3. Therefore, $f_\mu \notin B_m \cap \mathrm{T}(M_2)$.■

4. Cocharacters, Hilbert Series and Codimensions

Theorem 4.1. [9, 5] *The cocharacter of the T-ideal* $\mathrm{T}(M_2)$ *is* $\chi_n(M_2) = \sum_{\lambda\vdash n} \kappa(\lambda)\chi(\lambda)$, $n = 0,1,2,\dots,$ *where* $\lambda = (\lambda_1,\lambda_2,\lambda_3,\lambda_4)$ *and*

(i) $\kappa(n) = 1$;

(ii) $\kappa(\lambda_1,\lambda_2) = (\lambda_1 - \lambda_2 + 1)\lambda_2$, *if* $\lambda_2 > 0$;

(iii) $\kappa(\lambda_1,1,1,\lambda_4) = (\lambda_1 + 1)(2 - \lambda_4) - 1$;

(iv) $\kappa(\lambda_1,\lambda_2,\lambda_3,\lambda_4) = (\lambda_1 - \lambda_2 + 1)(\lambda_2 - \lambda_3 + 1)(\lambda_3 - \lambda_4 + 1)$

for all other partitions.

Proof. By the similarity between the S_n-modules $P_n/(P_n \cap \mathrm{T}(M_2))$ and GL_m-module $F_m(\mathrm{var}\ M_2)$, it suffices to obtain the decomposition of $F_m(\mathrm{var}\ M_2)$. In virtue of Theorem 2.2 (iii) and Theorem 3.4

$$F_m(\mathrm{var}\ M_2) = K[x_1,\dots,x_m] \otimes \left(\sum N_m(\mu_1,\mu_2,\mu_3)/ (\sum_{n\geq 1} N_m(n) + N_m(1^3))\right).$$

Following Theorem 2.2 (iv) and the example below Proposition 1.1, we see that all irreducible components of $F_m(\mathrm{var}\ M_2)$ correspond to Young diagrams with at most four rows. Let $N_m(\lambda)$, $\lambda = (\lambda_1,\lambda_2,\lambda_3,\lambda_4)$ participate in the tensor product

$$U_1 = K[x_1,\dots,x_m] \otimes \sum N_m(\mu_1,\mu_2,\mu_3).$$

Then $[\lambda]$ can be obtained from certain $[\mu] = [\mu_1,\mu_2,\mu_3]$ by adding $0,1,\dots,\lambda_1 - \lambda_2$ shaded boxes to the first row of $[\mu]$, $0,1,\dots,\lambda_2 - \lambda_3$ ones to the second and $0,1,\dots,\lambda_3 - \lambda_4$ - to the third row.

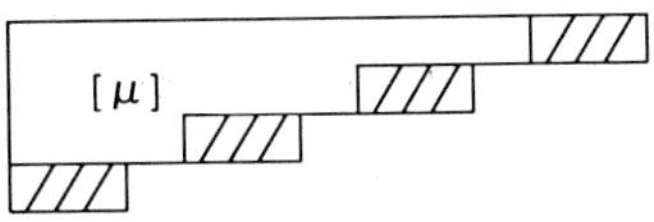

Therefore, $N_m(\lambda)$ appears in the tensor product U_1 with multiplicity $(\lambda_1 - \lambda_2 + 1)(\lambda_2 - \lambda_3 + 1)(\lambda_3 - \lambda_4 +1)$. Similar considerations show that

$$U_2 = K[x_1,\dots,x_m] \otimes \Big(\sum_{n\geq 1} N_m(n) + N_m(1^3)\Big) = \sum \varphi(\lambda) N_m(\lambda) =$$

$$=\Big(K[x_1,\dots,x_m] \otimes (K[x_1,\dots,x_m] + N_m(1^3))\Big)/K[x_1,\dots,x_m] =$$

$$= \sum (\lambda_1 - \lambda_2 + 1) N_m(\lambda_1,\lambda_2) + \sum N_m(\lambda_1,1^2) + \sum N_m(\lambda_1,1^3) - \sum_{n\geq 0} N_m(n).$$

Clearly, $\varphi(\lambda) \neq 0$ only in the cases (i), (ii) and (iii). Since

$$F_m(\text{var } M_2) = \sum\Big((\lambda_1 - \lambda_2 + 1)(\lambda_2 - \lambda_3 + 1)(\lambda_3 - \lambda_4 +1) - \varphi(\lambda)\Big) N_m(\lambda)$$

easy calculations complete the proof.∎

Theorem 4.2. [15, 9, 5] *The Hilbert series of* $F_m(\text{var } M_2)$ *satisfies:*

$$H(F_m(\text{var } M_2), t_1, \dots, t_m) = \prod_{i=1}^{m} (1 - t_i)^{-1} \left(\sum_{\mu} S_{(\mu_1, \mu_2, \mu_3)}(t_1, \dots, t_m) + \right.$$

$$\left. + 1 - \prod_{i=1}^{m} (1 - t_i)^{-1} - e_3(t_1, \dots, t_m) \right) =$$

$$= \prod_{i=1}^{m} (1 - t_i)^{-2} \left(\sum_{\ell} S_{(\ell^2)}(t_1, \dots, t_m) - 1 \right) +$$

$$+ \prod_{i=1}^{m} (1 - t_i)^{-1} \left(1 - e_3(t_1, \dots, t_m) \right).$$

where $e_3(t_1, \dots, t_m)$ *is the third elementary symmetric function.*

Proof. The first equality follows immediately from Theorem 2.2 (ii) and Theorem 3.4. Now, Proposition 1.1 gives

$$\sum N_m(\ell^2) \otimes \sum N_m(n) = \sum N_m(\mu_1, \mu_2, \mu_3)$$

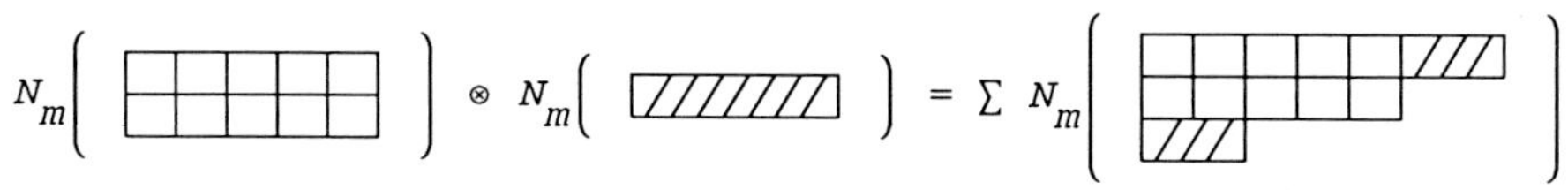

and the proof is completed by Theorem 3.2.∎

For $m \leq 4$ the Hilbert series of $F_m(\text{var } M_2)$ has an especially simple form.

Theorem 4.3. [9]

$$H(F_4(\text{var } M_2), t_1, t_2, t_3, t_4) = \prod_{i=1}^{4} \frac{1}{(1 - t_i)^2} \left(\frac{1 - t_1 t_2 t_3 t_4}{\prod_{i<j} (1 - t_i t_j)} - 1 \right) +$$

$$+ \prod_{i=1}^{4} \frac{1}{1 - t_i} \left(1 - e_3(t_1, t_2, t_3, t_4) \right).$$

Proof. The result of Thrall mentioned in Section 1 gives

$$N_4(1^2)^{\otimes^s k} = \sum N_4(\lambda_1^2, \lambda_2^2), \quad \lambda = (\lambda_1, \lambda_2) \vdash k,$$

for $m = 4$ because $N_m(\mu_1, \dots, \mu_r) = 0$ when $\mu_r \neq 0$ and $r > 4$. By the Littlewood-Richardson rule (see [13, Ch.1])

$$N_4(\lambda_1^2,\lambda_2^2) = N_4(\lambda_2^4) \otimes N_4((\lambda_1 - \lambda_2)^2).$$

Restated for the Schur functions this equality gives that the Hilbert series of $\sum N_4(1^2)^{\otimes^s k}$ equals

$$\sum S_{(p^4)}(t_1,t_2,t_3,t_4) \sum S_{(q^2)}(t_1,t_2,t_3,t_4) =$$

$$= \prod_{i=1}^{4} \frac{1}{1 - t_1t_2t_3t_4} \sum S_{(q^2)}(t_1,t_2,t_3,t_4).$$

Since the GL_4-module $\sum N_4(1^2)^{\otimes^s k}$ is isomorphic to the symmetric algebra of $N_4(1^2)$ and as a vector space $N_4(1^2)$ has a basis x_ix_j, $1 \le i < j \le 4$,

$$H(\sum N_4(1^2)^{\otimes^s k}, t_1,t_2,t_3,t_4) = \frac{1 - t_1t_2t_3t_4}{\prod_{i<j}(1 - t_it_j)}$$

and the proof follows from Theorem 4.2.∎

Theorem 4.4. [15] (see also [6])

(i) $$c(M_2,t) = \sum c_n(M_2)t^n = \frac{1}{t^2}(1 - 2t - \sqrt{1-4t}) - $$

$$- \frac{t^3}{(1-t)^4} + \frac{1}{1-t} - \frac{1}{1-2t};$$

(ii) $$c_n(M_2)t^n = \frac{1}{n+2}\binom{2n+2}{n+1} - \binom{n}{3} + 1 - 2^n;$$

(iii) (See [19]) *The n-th codimension* $c_n(M_2)t^n$ *is asymptotically equal to* $\dfrac{4^{n+1}}{n\sqrt{\pi n}}$.

Proof. (i) Applying Theorem 2.3 (iii) and Theorem 3.4, it suffices to compute the proper codimension series

$$\gamma(M_2,t) = \sum \gamma_k(M_2)t^k = \sum_k \sum_{\mu\vdash k} \dim M(\mu_1,\mu_2,\mu_3)t^k -$$

$$- \sum_{k\ge1} \dim M(k)t^k - \dim M(1^3)t^3.$$

Let us consider a hypotetical PI-algebra R such that

$$B_m/(B_m \cap T(R)) = \sum_{\ell\geq 0} N_m(\ell^2).$$

As in the proof of Theorem 4.2,

$$F_m(\text{var } R) = \sum_k \sum_{\mu\vdash k} N_m(\mu_1,\mu_2,\mu_3).$$

In virtue of Theorem 2.3 (iii)

$$c(R,t) = \frac{1}{1-t}\,\gamma(R,\frac{t}{1-t}) \quad\text{and}\quad \gamma(R,t) = \sum_{\ell\geq 0} \dim M(\ell^2)\,t^{2\ell}.$$

Applying the hook formula, we obtain $\dim M(\ell^2) = \frac{1}{\ell+1}\binom{2\ell}{\ell}$ and, as a consequence,

$$\gamma(R,t) = \sum \binom{2\ell}{\ell}\frac{t^{2\ell}}{\ell+1} = \frac{1}{2t^2}\left(1-\sqrt{1-4t^2}\right).$$

Easy calculations complete the proof, bearing in mind that

$$\sum_{k\geq 1} \dim M(k)\,t^k + \dim M(1^3)\,t^3 = \frac{1}{1-t} - 1 + t^3.$$

The assertion (ii) is equivalent to (i) and (iii) follows from the Stirling formula $n! \approx \sqrt{2\pi n}\,(\frac{n}{e})^n$.■

One of the most intriguing problems for a PI-algebra R is to find a generating set of $T(R)$ (or a **basis** of the polynomial identities for R). Razmyslov [16] proved that all polynomial identities for M_2 are consequences of these of degree ≤ 6 and found a basis of nine identities. There were several efforts to simplify his basis and the final result was given by the author [3].

Theorem 4.5. [16, 3] *As a T-ideal* $T(M_2)$ *is generated by the standard polynomial* $s_4(x_1,x_2,x_3,x_4)$ *and the Hall polynomial (in two variables)* $[[x_1,x_2]^2,x_1]$.

Idea of the proof. Let U be the T-ideal of $K\langle X\rangle$ generated by the polynomials $s_4(x_1,x_2,x_3,x_4)$ and $[[x_1,x_2]^2,x_1]$. Since the 2×2 matrix algebra satisfies both the standard and Hall identities, $U \subseteq T(M_2)$. Hence $B_m^{(n)} \cap U \subseteq B_m^{(n)} \cap T(M_2)$, where $B_m^{(n)}$ is the homogeneous component of degree n of B_m. By the result of Razmyslov [16] the theorem will be proved if we show that $B_m^{(n)} \cap U = B_m^{(n)} \cap T(M_2)$ for all $n \leq 6$. Practically the cases $n = 5$ and $n = 6$ have to be considered only. The GL_m-module structure of $B_m^{(5)}$ and $B_m^{(6)}$ was obtained in [2, 4] and this together with Theorem 3.4 gives the description of $B_m^{(n)} \cap T(M_2)$ for $n = 5$ and $n = 6$. Straightforward calculations show that the multiplicities of the irreducible components of the GL_m-modules $B_m^{(n)} \cap U$ and $B_m^{(n)} \cap T(M_2)$ are the same for $n = 5$ and $n = 6$ and we do have the equality in $B_m^{(n)} \cap U \subseteq B_m^{(n)} \cap T(M_2)$, $n = 5, 6$.■

Remark 4.6. *Let* $\overline{F}_m$ *be the* 2×2 *generic trace algebra generated by the generic* 2×2 *matrix algebra* $<y_1, \dots, y_m>$ *and the traces of its elements.* Procesi [15] showed that *the algebras* $\overline{F}_m$ *and* $K[\mathrm{tr}\, y_1, \dots, \mathrm{tr}\, y_m] \otimes K<x_1, \dots, x_m>/W$ *are isomorphic, where* W *is the ideal of* $K<x_1, \dots, x_m>$ *consisting of all polynomials* $f(x_1, \dots, x_m)$ *such that* $f(g_1, \dots, g_m) = 0$ *for all* $g_1, \dots, g_m \in sl_2 \subset M_2$. (These polynomials are called **weak identities for the pair** (M_2, sl_2)). Using invariant theory Procesi [15, 12, Ch.4] proved that

$$K<x_1, \dots, x_m>/W = \sum N_m(\mu_1, \mu_2, \mu_3) = K[x_1, \dots, x_m] \otimes \sum_{\ell \geq 0} N_m(\ell^2)$$

and mentioned that *the* GL_m*-module* $\sum_{\ell \geq 0} N_m(\ell^2)$ *is isomorphic to the homogeneous coordinate ring of the Grassmann variety of the 2-dimensional planes in the* m*-dimensional affine space (or, equivalently, of lines in* $\mathbb{P}^{m-1}$). The main step of his approach was the proof of the isomorphism $K<x_1,\ldots,x_m>/W = \sum N_m(\mu_1,\mu_2,\mu_3)$. Repeating verbatim the proof of Theorem 3.4 we can obtain a new proof of this result (see [7] for details).■

References

1. **A.Berele**, Homogeneous polynomial identities, *Israel J. Math* **42** (1982), 258 - 272.
2. **V.S.Drenski**, Representations of the symmetric group and varieties of linear algebras, *Mat.Sb.* **115** (1981), 98 - 115 (Russian). Translation: *Math. USSR Sb.* **43** (1981), 85 - 101.
3. **V.S.Drenski**, A minimal basis for the identities of a second-order matrix algebra over a field of characteristic 0, *Algebra i Logika* **20** (1981), 282 - 290 (Russian). Translation: *Algebra and Logic* **20** (1981), 188 - 194.
4. **V.S.Drenski**, Lattices of varieties of associative algebras, *Serdica* **8** (1982), 20 - 31 (Russian).
5. **V.Drensky**, Codimensions of T-ideals and Hilbert series of relatively free algebras, *J.Algebra* **91** (1984), 1 - 17.
6. **V.S.Drensky**, Explicit codimension formulas of certain T-ideals, *Sib.Mat.Zh.* **29** (1988) No.6, 30 - 36 (Russian). Translation: *Siberian Math.J.* **29** (1988), 897 - 902.
7. **V.Drensky**, Computational techniques for PI-algebras, in "Topics in Algebra", part 1, "Rings and Representations of Algebras", *Banach Center Publ.* **26**, Polish Scientific Publishers, Warsaw, 1990, 17 - 44.
8. **E.Formanek**, The polynomial identities of matrices, *Contemp.Math.* **13** (1982), 41 - 79.
9. **E.Formanek**, Invariants and the ring of generic matrices, *J.Algebra* **89** (1984), 178 - 223.
10. **E.Formanek, P.Halpin, W.-C.W.Li**, The Poincaré series of the ring of 2×2 generic matrices, *J.Algebra* **69** (1981), 105 - 112.
11. **G.James, A.Kerber**, "The Representation Theory of the Symmetric Group", Encyclopedia of Math. and Its Appl. **16**, Addison-Wesley, Reading, Mass., 1981.

12. **L. Le Bruyn**, Trace rings of generic 2 by 2 matrices, *Mem.Amer.Math.Soc* **66** (1987), No. 363.

13. **I.G.Macdonald**, "Symmetric Functions and Hall Polynomials", Oxford Univ. Press (Clarendon), Oxford, 1979.

14. **C.Procesi**, The invariant theory of $n \times n$ matrices, *Advances in Math.* 19 (1976), 306 - 381.

15. **C.Procesi**, Computing with 2×2 matrices, *J.Algebra* **87** (1984), 342 - 359.

16. **Ju.P.Razmyslov**, Finite basing of the identities of a matrix algebra of second order over a field of characteristic 0, *Algebra i Logika* **12** (1973), 83 - 113 (Russian). Translation: *Algebra and Logic* **12** (1973), 43 - 63.

17. **Yu.P.Razmyslov**, Trace identities of full matrix algebras over a field of characteristic zero, *Izv. AN SSSR, Ser. Mat.* **38** (1974), 723 - 756 (Russian). Translation: *Math.USSR Izv.* **8** (1974), 727 - 760.

18. **A.Regev**, Existence of identities in $A \otimes B$, *Israel J.Math.* **11** (1972), 131 - 152.

19. **A.Regev**, On the codimensions of matrix algebras, in "Algebra - Some Recent Trends", *Lect.Notes in Math.* **1352** , Springer Verlag, Berlin-Heidelberg-New York, 1988, 162 - 172.

20. **L.H.Rowen**, "Polynomial Identities in Ring Theory", Acad.Press, New York, 1979.

21. **W.Specht**, Gesetze in Ringen. I, *Math.Z.* **52** (1950), 557 - 589.

22. **M.R.Thrall**, On symmetrized Kronecker powers and the structure of the free Lie ring, *Amer.J.Math.* **64** (1942), 371 - 388.

23. **H.Weyl**, "The Classical Groups, Their Invariants and Representations", Princeton Univ. Press, Princeton, N.J., 1946.

(Received: October 1990)

Acta Applicandae Mathematicae **21**: 163–184, 1990.

CAYLEY FACTORIZATION AND A STRAIGHTENING ALGORITHM

Three talks presented to the Computational Algebra Seminar

Department of Mathematics, University of Rome "Tor Vergata"

May 9-11, 1990

by Neil L. White

Department of Mathematics, University of Florida, Gainesville, U.S.A.

1. THE CAYLEY FACTORIZATION PROBLEM AND ITS APPLICATIONS

The Cayley Algebra

An important problem in computer-aided geometric reasoning is to automatically find geometric interpretations for algebraic expressions. For projective geometry this question can be reduced to the Cayley factorization problem. Before describing this problem, we give a brief introduction to the Cayley algebra for those readers not already familiar with it. The Cayley algebra is essentially the same as the classical Grassmann algebra. This algebra provides projectively invariant algebraic interpretations of synthetic geometric statements. A more thorough exposition may be found in Barnabei *et al.* (1985), Doubilet *et al.* (1974), or Rota & Stein (1976).

Let V be a vector space of dimension d over a field F of characteristic zero. Let $\Lambda(V)$ denote the exterior algebra of V We will write the exterior product in $\Lambda(V)$ as $\vee$ rather than the usual $\wedge$, and refer to it as the *join* operation. As is well known, this product is associative, distributive over addition, and anti-symmetric.

Let $a_1, \ldots, a_k \in V$, and write $A = a_1 \vee a_2 \vee \ldots \vee a_k$ or $A = a_1 a_2 \cdots a_k$ for the join of the k vectors. Then A is called an *extensor of step* k, or a *decomposable* k-*vector*.

Let $B = b_1 b_2 \cdots b_j$ be an extensor of step j. Then

$$A \vee B = a_1 \vee a_2 \vee \ldots \vee a_k \vee b_1 \vee \ldots \vee b_j = a_1 a_2 \cdots a_k b_1 \cdots b_j$$

is an extensor of step $j + k$. In fact, $A \vee B$ is non-zero if and only if $a_1, a_2, \ldots, a_k$, $b_1, \ldots, b_j$ are distinct and linearly independent. If $\overline{A}$ denotes the span of

AMS subject classifications (1980) 05Axx, 15A75.
Key words. Dotted straightening algorithm, Caley factorization.

$\{a_1, a_2, \ldots, a_k\}$ when $A \neq 0$, then $\overline{A}$ is called the *support* of A. We note that A is determined up to scalar multiple by $\overline{A}$, a well-known fact from exterior algebra. If $a_1, a_2, \ldots, a_k, b_1, \ldots, b_j$ are distinct and linearly independent, then $\overline{A \vee B} = \overline{A} + \overline{B}$, and thus the join operation on extensors corresponds to the lattice join of subspaces of V in the independent case. It is for this reason that we have chosen to call the exterior product the "join" and we will shortly define a second operation which corresponds similarly to the lattice meet.

An extensor of step d is, up to a fixed non-zero multiple, the determinant of its d constituent vectors. We denote this multiple of the determinant by brackets, $A = a_1 a_2 \cdots a_d = [a_1, a_2, \ldots, a_d]$.

We now define the *meet* operation. If $A = a_1 a_2 \cdots a_j$ and $B = b_1 b_2 \cdots b_k$, with $j + k \geq d$, then

$$A \wedge B = \sum_{\sigma} \text{sign } (\sigma)[a_{\sigma(1)}, \ldots, a_{\sigma(d-k)}, b_1, \ldots, b_k] a_{\sigma(d-k+1)} \cdots a_{\sigma(j)}.$$

The sum is taken over all permutations σ of $\{1, 2, \ldots, j\}$ such that $\sigma(1) < \sigma(2) < \ldots < \sigma(d-k)$ and $\sigma(d-k+1) < \sigma(d-k+2) < \ldots < \sigma(j)$. Such permutations are called *shuffles* of the $(d-k, j-(d-k))$ split of A.

An alternate notation for such signed sums over shuffles is the Scottish (named for Turnbull, 1928) or *dotted* notation, which we will frequently employ. We simply place dots over the shuffled vectors, with the summation and sign (σ) implicit. Similarly, shuffles may be defined over splits into any number of parts, and denoted by dots. In this paper, the brackets are always delimiters which define the parts of the split. If we wish to sum over several shuffles of disjoint sets, we will use separate symbols (triangle, square) over the vectors of each shuffled set. Thus

$$A \wedge B = [\dot{a}_1, \ldots, \dot{a}_{d-k}, b_1, \ldots, b_k] \dot{a}_{d-k+1} \cdots \dot{a}_j.$$

If $j + k = d$ then $A \wedge B = [a_1, \ldots, a_j, b_1, \ldots, b_k]$. This is a scalar of step 0, and we must be careful to distinguish it from $A \vee B$, a scalar with the same numerical value but of step d.

The meet is dual to the join, where duality exchanges vectors with covectors (extensors of step $d-1$). The meet corresponds to lattice meet of subspaces, $\overline{A \wedge B} = \overline{A} \cap \overline{B}$, provided $\overline{A} \cup \overline{B}$ spans V. Thus the meet operation corresponds to our geometric intuition in the case that A and B themselves are non-degenerate and that $\overline{A} \cap \overline{B}$ has as small dimension as possible. The *Cayley algebra* is the vector space $\Lambda(V)$ together with the operations $\vee$ and $\wedge$.

We now illustrate the translation of geometric incidences into the Cayley algebra. Consider the affine plane over F. By its usual embedding into a projective plane, we have a vector space V of dimension 3 over F, with its subspaces of dimensions 1 and 2 corresponding to points and lines, respectively. Thus we can represent both affine and projective incidences in a symbolic, coordinate-free way in the Cayley algebra.

EXAMPLE 1. Consider 3 coincident lines, $\overline{ab}, \overline{cd}, \overline{ef}$.

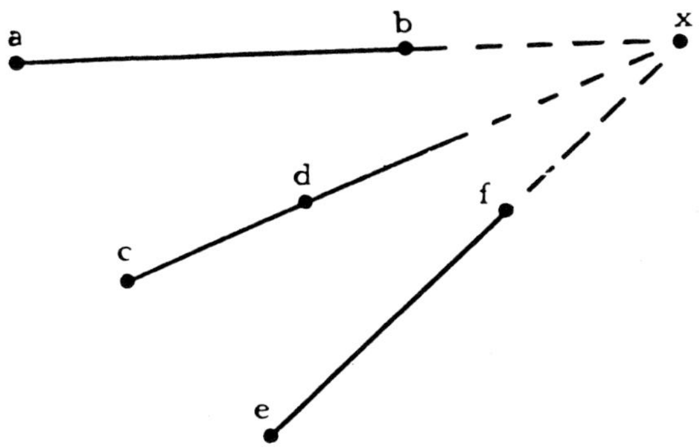

Then $ab \wedge cd = \alpha x$, where α is a scalar, and $x \wedge ef = 0$. Thus

$$0 = (ab \wedge cd) \wedge ef = ([acd]b - [bcd]a) \wedge ef = ([\dot{a}cd]\dot{b}) \wedge ef = [\dot{a}cd][\dot{b}ef].$$

In similar fashion, any incidence theorem or incidence relation in projective geometry may be translated into a conjunction of Cayley algebra statements, and, conversely, Cayley algebra statements may be translated back to projective geometry just as easily, provided they involve only join and meet, not addition. We will refer to Cayley algebra statements involving only join and meet as *simple* Cayley algebra statements. Now, simple Cayley algebra statements may be expanded into bracket statements by the definitions and properties of join and meet. Most Cayley algebra statements resulting from geometric incidence relations have step 0 or d. If a simple Cayley algebra statement $C(a, b, \dots) = 0$ has $C(a, b, \dots)$ of step k, $k \neq 0$ or d, then this is the equivalent of

$$C(a, b, \dots) \vee x_1 \vee \dots \vee x_{d-k} = [C(a, b, \dots), x_1, \dots, x_{d-k}] = 0$$

for all $x_1, \dots, x_{d-k} \in V$. This in turn is equivalent to $[C(a, b, \dots), y_1, \dots, y_{d-k}] = 0$ for all $y_1, \dots, y_{d-k} \in Y$ where Y is a basis of V. Thus every simple Cayley algebra statement is equivalent to a finite conjunction of bracket statements. However, the converse problem, that of writing a bracket statement as a simple Cayley algebra statement, when possible, is not easy. This is the problem we refer to as *Cayley factorization.*

Cayley Factorization

To explain the importance of this problem, let us introduce one more step away from the geometry, namely, the introduction of coordinates. Until now, we have been dealing with invariant languages with respect to the projective general linear group. Our Cayley algebra and bracket algebra statements may be expanded in terms of the coordinates of the vectors, but then we have statements in a larger algebra which includes non-invariant expressions. We may represent the situation in the following diagram (see also Example 2) :

(1)	Projective geometry	
	$\updownarrow$	
(2)	Cayley algebra	
	$\downarrow$	$\uparrow$ Cayley factorization
(3)	Bracket algebra	
	$\downarrow$	
(4)	Coordinate algebra	

Consider now the situation in computer-aided geometric reasoning. Considerable success has already been attained by automated geometry-theorem-proving programs in proving theorems in projective and Euclidean geometry, by going directly to the coordinate algebra (Chou *et al.*, 1987 and Kutzler & Stifter, 1986). Suppose, however, that we wish to use the computer in a more interactive fashion, where the computer reduces the problem to an algebraic one, does some symbolic manipulation, and then we wish to interpret these algebraic results geometrically in order to decide how to proceed. From ordinary coordinate algebra, it is computationally much more difficult to find such geometric interpretation in general than it is from the bracket algebra (see Example 2). Indeed, any such interpretation program would at some point have to ascend to invariant statements, and therefore contain in some disguised form a Cayley factorization algorithm. However, in projective synthetic geometry, we can do our symbolic computation directly in the bracket algebra, since brackets are projective invariants, and if we have a Cayley factorization algorithm, then we can more easily find such geometric interpretation when it exists.

There does not exist a usable Cayley factorization algorithm, but in an important special case, that in which the bracket condition is multilinear, we have an algorithm which works well for instances of up to about 16 points. This will be the subject of our second lecture.

EXAMPLE 2. We now give an example that illustrates the above translations, and also indicates some possibility of direct application of the Cayley factorization algorithm to geometry theorem proving. The numbering of statements corresponds to the numbering in the above diagram.

PAPPUS' THEOREM. *(1) If aa', bb', and cc' are concurrent lines, and ab', bc', ca' likewise concur, then ac', ba', cb' concur.*

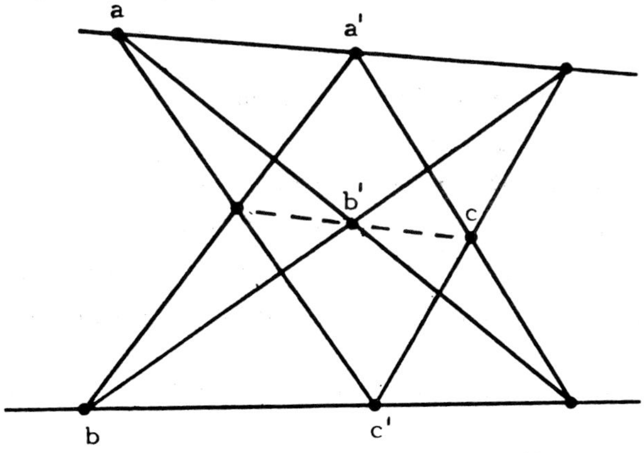

Let

$$P = aa' \wedge bb' \wedge cc' + ab' \wedge bc' \wedge ca'. \quad (2)$$

If the hypotheses of Pappus' Theorem are true, then $P = 0$. This translation of the projective geometry statements (hypotheses) into a Cayley algebra condition is trivial, except for the addition operation. In principle, any linear combination of the two simple Cayley algebra conditions might have sufficed to represent their conjunction, but only this particular linear combination will have the desired Cayley factorization (corresponding to the conclusion of the Theorem). Incidentally, it is not obvious why there should be one "right" translation in this sense, and this question deserves further study.

Expanding P as a bracket polynomial, we get (using anti-commutativity of $\wedge$)

$$P = [a, a', b'][b, c, c'] - [a, a', b][b', c, c'] + [a, b', c'][b, c, a'] - [a, b', b][c', c, a']. \quad (3)$$

It is not immediately obvious, but P may be Cayley factored as $-ac' \wedge ba' \wedge cb'$, thus giving a proof of Pappus' Theorem.

We may also expand P in the coordinate algebra, getting

$$P = \sum a_i a'_j b'_k b_p c_q c'_r - a_i a'_j b_k b'_p c_q c'_r + a_i b'_j c'_k b_p c_q a'_r - a_i b'_j b_k c'_p c_q a'_r, \quad (4)$$

where the sum is over all pairs of permutations (i, j, k) and (p, q, r) of $\{1, 2, 3\}$. This expression may be immensely simplified by choosing a basis, for example, $a' = e_1$, $b' = e_2$, $c' = e_3$. Then

$$\begin{aligned} P &= a_3b_1c_2 - a_3b_2c_1 - a_2b_3c_1 + a_3b_2c_1 + a_1b_2c_3 - a_1b_3c_2 + a_1b_3c_2 - a_3b_1c_2 \\ &= -a_2b_3c_1 + a_1b_2c_3. \end{aligned} \quad (4')$$

This expression is much simpler and more elementary (in that it doesn't require brackets, Cayley operations, or other esoterica). But can we recover the geometric meaning from this expression? Since "geometric meaning" should not depend on choice of basis, this means we must return to an expression in an invariant language, namely brackets. This can be done, but at the cost of getting a much higher degree bracket polynomial than previously. For our example, we have

$$P = -[e_1, a, e_3][e_1, e_2, b][c, e_2, e_3] + [a, e_2, e_3][e_1, b, e_3][e_1, e_2, c]. \quad (3')$$

This can then be Cayley factored as

$$P = -[e_1, e_2, e_3]\, ae_3 \wedge be_1 \wedge ce_2, \quad (2')$$

equivalent to the result we had previously. For an example of higher degree or higher rank initially, the increase in degree would be considerably greater. This increase in degree will in general make Cayley factorization far more difficult. Thus it is our claim that when geometric interpretation is desired, it is much better to remain at the bracket level, rather than to descend to coordinates. The bracket algebra is, in a very real sense, closer to the synthetic geometry.

Some more immediate applications of Cayley factorization are in structural rigidity (White & Whiteley, 1983 and 1987), convex polytopes (Bokowski & Sturmfels, 1987), and splines(Whiteley, 1990). The infinitesimal rigidity of certain kinds of frameworks, the dimension of the space of splines over a planar polyhedral complex, and the realizability of certain convex polytopes have been characterized in terms of bracket statements. Cayley factorization of these statements would provide direct geometric interpretation of these conditions.

EXAMPLE 3. We give an example of a bar-and-body framework, as shown in the figure. This framework consists of three rigid planar bodies, free to move in the Euclidean plane, and connected by six rigid bars, with the connections at the ends of each bar allowing free rotation of the bar relative to the rigid body.

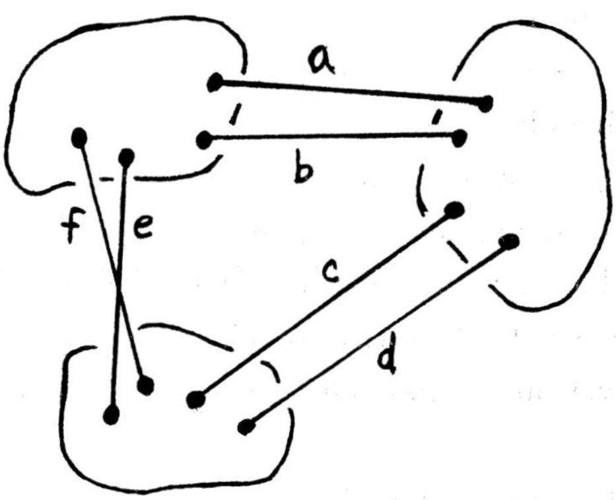

Let $V = I\!R^3$, and $W = \Lambda^2(V) \cong I\!R^3$. We think of the endpoints of the bars as elements of V, and hence the lines determined by the bars are two-extensors of these points, or elements of W. In White and Whiteley (1987), a direct combinatorial algorithm is given which provides the bracket condition for such a framework to have an infinitesimal motion. It consists of considering the framework as a graph, with the bodies becoming vertices and the bars edges, and writing one bracket term for each way of directing the edges so that a fixed body has outdegree zero and all other bodies have outdegree three. The bars directed out of a body become the bars in a bracket, and the relative signs between bracket terms are computed by the sign of the permutaton of the bars between the terms. In our example, we find that the framework has an infinitesimal motion if and only if $[abc][def]-[abd][cef] = 0$. Now, this bracket polynomial may be Cayley factored as $ab \wedge cd \wedge ef = 0$. Now we switch to thinking of $a, b, \ldots, f$ as 2-extensors in V rather than elements of W, and recall that we have duality between V and W, hence between $\Lambda(V)$ and $\Lambda(W)$. Thus the above becomes $(a \wedge b) \vee (c \wedge d) \vee (e \wedge f) = 0$. This is the desired geometric condition for the existence of an infinitesimal motion, and is in this case just an application of the classical theorem of Arnhold-Kempe that the centers of relative motion of three pairs of bodies must be collinear.

2. MULTILINEAR CAYLEY FACTORIZATION

In this lecture we will make repeated use of the straightening algorithm of classical invariant theory. This algorithm reduces a bracket polynomial, modulo the Grassmann-Plücker relations, to a canonical form which is a linear combination of standard tableaux. This may also be regarded as a normal form algorithm with respect to a Gröbner basis. In particular, it is necessary in order to decide if a bracket polynomial is equal to zero or not. We may also make use of a variation of the straightening algorithm, called the *dotted* straightening algorithm. This algorithm is the subject of our third lecture, but for now we will simply point out that it is normal straightening applied to bracket polynomials which have some dotting, and that it preserves this dotting. It greatly improves the efficiency of the multilinear Cayley factorization algorithm which is the subject of the present lecture. However, all of these algorithms are exponential in time and space complexity.

Let $P(a, b, c, \ldots)$ be a bracket polynomial which is the expansion of a simple Cayley algebra expression. Then P is homogeneous, that is, for each (variable) vector x which is an argument of P, there is a positive integer n_x such that each monomial of P has precisely n_x occurrences of x among the vectors in its brackets. In fact, n_x is also the total number of occurrences of x in the simple Cayley algebra expression. If $n_x = 1$ for all x, then P is *multilinear*. The bracket condition which arose in Example 3 is multilinear, as it always is for a bar-and-body condition.

but that in Example 2 is not. Now suppose that $ab\cdots e$ occurs explicitly as a join of vectors in the simple Cayley expression. We will refer to $ab\cdots e$ as an *atomic extensor* in that case. Each atomic extensor appears in P as a dotted set of symbols, where undotted symbols all in the same bracket may be considered a special case of being dotted.

EXAMPLE 4. Let $d = 5$. Then

$$(abcd) \wedge (efgh) \wedge (amn) \wedge (bekl) = ([\dot{a}efgh]\dot{b}\dot{c}\dot{d}) \wedge (amn) \wedge (bekl)$$

$$= ([\dot{a}efgh][\dot{b}\dot{c}amn]\dot{d}) \wedge (bekl) = [\dot{a}efgh][\dot{b}\dot{c}amn][\dot{d}bekl],$$

where the dotting here indicates the signed sum over all permutations $\sigma(a), \sigma(b), \sigma(c), \sigma(d)$ such that $\sigma(b) < \sigma(c)$ with ‘$<$’referring to alphabetical order. Note that some of these permutations cause repeated letters in a bracket, making those terms zero.

Let $C(a, b, c, \dots)$ be a simple Cayley expression which is multilinear in $a, b, c, \dots$. We may represent C by a binary tree T which has a point $a, b, c, \dots$ at each leaf, with distinct leaves having distinct points, and an operation $\vee$ or $\wedge$ at each inner node. The step of each subtree of T may be computed recursively from the facts that if A and B are extensors of steps j and k, respectively, then $A \vee B$ and $A \wedge B$ have steps $j + k$ and $j + k - d$, respectively. We will say that an operation $\vee$ or $\wedge$ is *trivial* if (1) its operands have steps j and $d - j$ for some j, or, (2) one of its operands has step 0 or d. We note that in case (1) a trivial $\vee$ may be replaced by a $\wedge$, with the resulting extensor having step 0 instead of step d, and vice-versa. If this occurs in a subtree, then other operations higher in the tree may also have to be changed to accommodate the changed steps, but it is not hard to see that this may be done without changing the geometric condition represented by C. In case (2), if B has step d, then $A \wedge B$ may replaced by $A \vee B$ by changing B to be of step 0 by changes as in case (1), and similarly for $A \vee B$ with B of step 0. In all cases, having a trivial operation amounts to having a subtree which evaluates to a bracket polynomial which factors out of C.

In the following, indeterminate points will mean points with algebraically independent coordinates. We also refer to this situation as generic position, and remark that it follows that the points are in general position in a strong sense, namely that no simple multilinear Cayley expression in such points can be zero, unless it is automatically zero by virtue of the step sizes.

We now consider several strong results about the structure that multilinear simple Cayley expressions can have, that lead directly to our algorithm for solving the multilinear Cayley factorization problem. The proofs of these results may be found in White (1991).

LEMMA 1. *Let $C(a, b, c, \dots)$ be a non-zero multilinear simple Cayley expression of step d in the indeterminate points $a, b, c, \dots$. Then C is anti-symmetric in two arguments a and b if and only if a and b do not have a non-trivial $\wedge$ on the unique path joining them in T.*

Now, if $P(a, b, c, \dots)$ is a multilinear bracket polynomial, we define $a \sim b$ if $a = b$ or if $P(a, b, \dots, z) = -P(b, a, \dots, z)$. We note that $\sim$ is an equivalence relation on the points.

COROLLARY 2. *If P is a multilinear bracket polynomial which has a Cayley factorization, then each equivalence class of $\sim$ is precisely the atomic extensor in some Cayley factorization of P.*

LEMMA 3. *Let P be a non-zero multilinear bracket polynomial for which there exists a Cayley factorization $P = C$, and let A and B be atomic extensors (i.e., equivalence classes of points) of P, and suppose that $P = (A \wedge B) \vee Q$ for some linear combination Q of extensors with bracket coefficients. Then there is no non-trivial join separating A and B in the tree T corresponding to C, and hence there exists a Cayley factorization $P = (A \wedge B) \vee U$ where U is a simple Cayley expression.*

The Multilinear Cayley Factorization Algorithm

Now we are ready to describe the Cayley factorization algorithm for multilinear bracket polynomials. Let $P(a, b, \dots, z)$ be a bracket polynomial of rank d which is multilinear in the N points $a, b, \dots, z$.

Step 1. We first find the atomic extensors. We have defined $a \sim b$ if $a = b$ or if $P(a, b, \dots, z) = -P(b, a, \dots, z)$. The latter is checked by applying the straightening algorithm to $P(a, b, \ldots, z) + P(b, a, \ldots, z)$, or, equivalently (in the multilinear case), to $P(a, a, \ldots, z)$, to see if zero is obtained. The transitivity of $\sim$ cuts down on the number of pairs of points we have to check, although it is still $O(N^2)$ calls of the straightening algorithm in the worst case. The equivalence classes are precisely the maximal atomic extensors, if a factorization exists, by Corollary 2. If there do not exist two atomic extensors whose sizes sum to at least d, then return NO FACTORIZATION POSSIBLE.

Step 2. Rewrite P as a bracket polynomial which is dotted in each atomic extensor. If $\{a, b, \ldots, e\} = E$ is an atomic extensor of m points, then

$$\begin{aligned} P(a, b, \ldots, e, f, \ldots, z) &= \frac{1}{m!} \sum_{\sigma} \text{sign}\,(\sigma) P(\sigma(a), \sigma(b), \ldots, \sigma(e), f, \ldots, z) \\ &= Q(\dot{a}, \dot{b}, \ldots, \dot{e}, f, \ldots, z) \end{aligned}$$

where the sum is over all permutations σ of $\{a, b, \ldots, e\}$, and antisymmetry of the brackets is used to sort the letters of E within each bracket, with terms then collected to form Q. This is repeated for each atomic extensor, and the disjointness of the atomic extensors assures that the dotting of earlier extensors is not spoiled. We now assume that Q is so dotted.

Step 3. If there exists an atomic extensor E of step d, then apply (dotted) straightening with the d elements of E first in the linear order. The result must have E as the first row of every resulting tableau, that is, $[E]$ is now an explicit factor, $Q = [E] \cdot Q'$. Remove E, store $Q = E \wedge Q'$, and proceed with the bracket polynomial Q' replacing Q. Repeat step 3 as appropriate. If $Q = 1$, then we are DONE, and the required Cayley factorization may be reconstructed.

Step 4. Find two atomic extensors $E = \{e_1, e_2, \ldots, e_k\}$ and $F = \{f_1, f_2, \ldots, f_\ell\}$ such that $E \wedge F$ could be a primitive factor, if they exist. We need only check pairs of extensors E, F such that $k + \ell \geq d$. By a primitive factor, we mean that $(e_1 e_2 \cdots e_k \wedge f_1 f_2 \cdots f_\ell)$ occurs explicitly in some Cayley factorization of Q. Note that any simple Cayley expression must have such a primitive factor. A necessary condition for $E \wedge F$ to be a primitive factor is that if the (dotted) straightening algorithm is applied to Q, using an ordering in which $e_1 < e_2 < \ldots < e_k < f_1 < f_2 < \ldots < f_\ell < x$ for all $x \in A - (E \cup F)$, then the result is in the form

$$\sum_{x_1,\ldots,y_1\ldots} \begin{array}{cccccccccc} e_1 & \cdots & \cdots & \cdots & \cdots & e_k & \dot{f}_1 & \cdots & \dot{f}_{d-k} \\ \dot{f}_{d-k+1} & \cdots & \dot{f}_\ell & x_1 & \cdots & \cdots & \cdots & \cdots & x_{2d-k-\ell} \\ y_1 & \cdots & \cdots & \cdots & \cdots & \cdots & \cdots & \cdots & y_d \\ \vdots & & & & \vdots & & & & \vdots \end{array} \qquad (*)$$

That is, every tableau has E in the first row, part of F filling up the rest of the first row, and the rest of F in the second row. The sum is over various terms with different choices for the x's and y's.

Actually, we are ignoring two global signs, since

$$\begin{aligned} E \wedge F = \pm F \wedge E &= \pm([\dot{f}_1, \ldots, \dot{f}_{d-k}, e_1, \ldots, e_k]\dot{f}_{d-k+1} \cdots \dot{f}_\ell) \\ &= \pm(\pm[e_1, \ldots, e_k, \dot{f}_1, \ldots, \dot{f}_{d-k}]\dot{f}_{d-k+1} \cdots \dot{f}_\ell), \end{aligned}$$

but this sign has no bearing on the existence of a factorization, and can be reconstructed if needed.

Step 5. If such E and F do not exist, then return NO FACTORIZATION POSSIBLE. If they do exist, then choose new symbols $g_1, g_2, \ldots, g_p$, where $p = k + \ell - d$, and store $G = E \wedge F$. Let G replace E and F in the collection of atomic extensors. Proceed with the new polynomial

$$P = \sum_{x_1,\ldots,y_1,\ldots} \begin{array}{cccccc} g_1 & \cdots & g_p & x_1 & \cdots & x_{2d-k-\ell} \\ y_1 & \cdots & \cdots & \cdots & \cdots & y_d \\ \vdots & & & \vdots & & \vdots \end{array}$$

where $x_1, \ldots, y_1, \ldots$ are the same as above. Now P has one less bracket per term than previously. Note that Lemma 3 guarantees that no backtracking will be necessary, since $E \wedge F$ must be part of a Cayley factorization if one exists.

Step 6. Recompute the atomic extensors by trying to extend the current ones.

Step 7. Go to Step 2.

This completes the algorithm. The correctness of the algorithm follows from the lemmas, as already indicated.

We now illustrate how the algorithm works.

EXAMPLE 5. Let

$$P(a,b,\dots,k) = 2[a,b,c][d,e,\overset{\blacktriangle}{f}][\overset{\blacktriangle}{g},h,k] - [\overset{\bullet}{a},c,e][\overset{\bullet}{b},d,\overset{\blacktriangle}{f}][\overset{\blacktriangle}{g},h,k] + [\overset{\bullet}{a},c,\overset{\blacktriangle}{f}][\overset{\bullet}{b},d,e][\overset{\blacktriangle}{g},h,k].$$

We have chosen an example of a multilinear bracket polynomial which is already in dotted form for the extensors ab, fg, and hk. Completing step 1, we find that each of these is maximal, but that cd is also an atomic extensor. Then

$$P = \frac{1}{2}P(\dots,c,d,\dots) - \frac{1}{2}P(\dots,d,c,\dots)$$

$$= [a,b,\overset{\blacksquare}{c}][\overset{\blacksquare}{d},e,\overset{\blacktriangle}{f}][\overset{\blacktriangle}{g},h,k] - \frac{1}{2}[\overset{\bullet}{a},\overset{\blacksquare}{c},e][\overset{\bullet}{b},\overset{\blacksquare}{d},\overset{\blacktriangle}{f}][\overset{\blacktriangle}{g},h,k] + \frac{1}{2}[\overset{\bullet}{a},\overset{\blacksquare}{c},\overset{\blacktriangle}{f}][\overset{\bullet}{b},\overset{\blacksquare}{d},e][\overset{\blacktriangle}{g},h,k].$$

Letting $E = ab$, $F = cd$, and applying dotted straightening with respect to alphabetical order, we find

$$P = [a,b,\overset{\blacksquare}{c}][\overset{\blacksquare}{d},e,\overset{\blacktriangle}{f}][\overset{\blacktriangle}{g},h,k].$$

Thus we have a single dotted tableau of the form required for step 5, as in (*),

$$P = \begin{matrix} a & b & \overset{\blacksquare}{c} \\ \overset{\blacksquare}{d} & e & \overset{\blacktriangle}{f} \\ \overset{\blacktriangle}{g} & h & k \end{matrix}$$

Letting $W = E \wedge F$, $W = w$ is an extensor of step 1, and

$$P = \begin{matrix} w & e & \overset{\blacktriangle}{f} \\ \overset{\blacktriangle}{g} & h & k \end{matrix}.$$

Now we recompute atomic extensors, obtaining we, fg, and hk, and we see that $we \wedge fg$ is already a factor. Letting $U = u = we \wedge fg$, $P = uhk$, and by step 3, we are done, as uhk is now an atomic extensor. Reconstructing,

$$P = uhk = u \wedge hk = (we \wedge fg) \wedge hk = (((ab \wedge cd) \vee e) \wedge fg) \wedge hk,$$

up to a possible minus sign.

We note that P is zero when the following incidences hold:

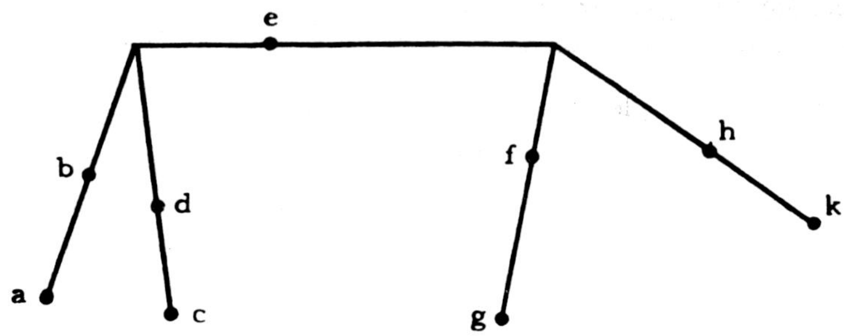

Clearly, several other simple Cayley expressions are equal to $\pm P$, for example, $(cd \wedge ((fg \wedge hk) \vee e)) \wedge ab$.

Implementation

The Cayley factorization algorithm, using only the ordinary straightening algorithm, has been implemented in FORTRAN on an Apollo 3000. As a test case, we obtained the bracket expansion of the simple rank 5 multilinear Cayley expression

$$(((cdf \wedge agh) \vee bij) \wedge klm) \wedge eno,$$

and modified the bracket polynomial by applying several syzygies to it. Using the resulting bracket polynomial as input, the program found a correct Cayley factorization in 2 minutes and 43 seconds, over 2 minutes of which was spent on Step 1. It was similarly successful on a number of other inputs, both factorable and non-factorable. However, 15 or 16 points is approximately the limit that this implementation can handle. The program is limited to 1500 tableaux during each call of the straightening algorithm, because of memory limitations of the Apollo, and this stack size has been exceeded for some inputs of 15 and of 16 points, after running for several hours, with most of the time spent in Step 4. An implementation using the dotted straightening algorithm would be much faster in time spent on Step 4, and would require smaller stack sizes, but would perhaps still be unable to handle inputs of more than about 25 points. FORTRAN source code may be obtained from the author.

Questions

Some questions remain. The first is to find an algorithm which can handle non-multilinear polynomials. This becomes more complicated, of course, and a straightforward modification of the multilinear algorithm will not suffice. We can attempt to define atomic extensors by $a \sim b$ if $P(b, b, c, \dots) = 0$, for example, but the results corresponding to Lemma 1, its corollary, and Lemma 3 are no longer true. Of course, there is a simple-minded algorithm which consists of listing all Cayley expressions on the same set of points with the appropriate multiplicities, and see if any expand to P. Thus the general Cayley factorization question is decidable.

The second concerns the fact that examples are known of non-Cayley-factorable bracket polynomials, which become Cayley factorable when multiplied by a single bracket. Sturmfels and Whiteley (1991) show that this always happens if we multiply by an appropriate product of brackets of high degree. Such extra bracket factors may correspond to the non-degeneracy factors that occur in Wu's method (see Chou *et al.*, 1987). How can such factors of small degree be found practically? Again, there exists a simple-minded algorithm, since Sturmfels and Whiteley provide a bound on the degree of the multiplier.

3. THE DOTTED STRAIGHTENING ALGORITHM

Dotted bracket expressions and compact tableaux

Dotted bracket expressions are bracket expressions, or vector invariants, which have additional anti-symmetry indicated on certain sets of points. In the previous two lectures we have seen dotted bracket expressions arise as the result of evaluating simple Cayley algebra expressions. They are also very useful in expressing invariants of anti-symmetric tensors or Cayley algebra extensors, as seen in McMillan (1990), an application which we will not explore in these lectures. We will represent dotted bracket expressions by simpler tableaux, which we call *compact tableaux*.We will obtain identities for these tableaux, which allow us to mimic the straightening algorithm. We will show that in the case of a dotted bracket expression, the classical straightening algorithm has been expressed by our new algorithm in a much more compact form, which can also be implemented in a much more efficient fashion. This compact straightening algorithm maintains the dotted form of the expression at every step, and achieves the same outcome as the ordinary straightening algorithm. We will restrict our attention to multilinear dotted bracket expressions.

Let P be a bracket product of degree m in brackets of length n, involving distinct points, that is dotted in the sets of points $a_1 a_2 \ldots a_{k_1}, b_1 b_2 \ldots b_{k_2}, \ldots, d_1 d_2 \ldots d_{k_j}$. We define the $m \times n$ tableau described below to be equal to P.

1) Rows of the tableau correspond to brackets of P, the first row to the first bracket, etc.

2) The entries of the rows will be the same letters as in the corresponding brackets of P without subscripts.

3) There is a sign attached, that being sign(σ) where σ is the permutation that takes the bracket entries in the order they are given in P and orders them lexicographically, $a_1, a_2, \ldots, a_{k_1}, b_1, b_2, \ldots, b_{k_2}, \ldots$

EXAMPLE.

$$[a_1 a_2 a_3 \overset{\bullet}{c_1}][\overset{\blacktriangle}{b_1} \overset{\blacktriangle}{b_2} \overset{\bullet}{c_2} \overset{\bullet}{c_3}][\overset{\blacktriangle}{b_3} d_1 d_2 d_3] = -\begin{pmatrix} a & a & a & c \\ b & b & c & c \\ b & d & d & d \end{pmatrix}.$$

Note that the above example is a signed sum of nine ordinary bracket monomials.

We have the negative sign since an odd permutation σ gives

$$\sigma\{a_1 a_2 a_3 c_1 b_1 b_2 c_2 c_3 b_3 d_1 d_2 d_3\} = \{a_1 a_2 a_3 b_1 b_2 b_3 c_1 c_2 c_3 d_1 d_2 d_3\}.$$

Note that the above compact tableau equals several different appearing versions of a dotted expression. For example:

$$\begin{pmatrix} a & a & a & c \\ b & b & c & c \\ b & d & d & d \end{pmatrix} = -[a_1 a_2 a_3 \overset{\bullet}{c_1}][\overset{\blacktriangle}{b_1} \overset{\blacktriangle}{b_2} \overset{\bullet}{c_2} \overset{\bullet}{c_3}][\overset{\blacktriangle}{b_3} d_1 d_3 d_3]$$

$$= [a_1 a_2 a_3 \overset{\bullet}{c_2}][\overset{\blacktriangle}{b_1} \overset{\blacktriangle}{b_2} \overset{\bullet}{c_1} \overset{\bullet}{c_3}][\overset{\blacktriangle}{b_3} d_1 d_2 d_3]$$

$$= [a_1 a_2 a_3 \overset{\bullet}{c_1}][\overset{\blacktriangle}{b_1} \overset{\blacktriangle}{b_3} \overset{\bullet}{c_2} \overset{\bullet}{c_3}][\overset{\blacktriangle}{b_2} d_1 d_2 d_3]$$

etc.

EXAMPLE. An example of a non-trivial dotted bracket expression is the superbracket of six pairs of points in rank four, which is the determinant of the Plücker coordinate vectors of the six lines determined by the pairs. This invariant is very important in structural rigidity, for if two rigid bodies are joined by six rigid bars attached at flexible endpoints, then the whole structure is infinitesimally rigid if and only if the superbracket of the six lines determined by the bars is non-zero. The superbracket yields the following tableaux in McMillan(1990):

$$\begin{pmatrix} a & a & d & e \\ b & b & d & f \\ c & c & e & f \end{pmatrix} - \begin{pmatrix} a & b & d & d \\ a & c & e & e \\ b & c & f & f \end{pmatrix}.$$

Each of these two tableaux represents a bracket monomial dotted in three pairs of points, and each is therefore equal to a signed sum of 8 ordinary bracket monomials. If the ordinary straightening algorithm is applied to this 16-term bracket polynomial, the result is again expressible as a dotted bracket expression, and hence as a linear combination of compact tableaux, namely

$$\begin{pmatrix} a & a & b & b \\ c & c & d & e \\ d & e & f & f \end{pmatrix} - \begin{pmatrix} a & a & b & c \\ b & c & d & d \\ e & e & f & f \end{pmatrix} - \begin{pmatrix} a & a & b & c \\ b & d & d & e \\ c & e & f & f \end{pmatrix} + \begin{pmatrix} a & a & b & d \\ b & c & c & e \\ d & e & f & f \end{pmatrix}.$$

Thus the dotted brackets impose an algebra on compact tableaux. In particular, we will see that the ordinary straightening algorithm applied to dotted bracket expressions induces a straightening algorithm on compact tableaux.

The entries of a compact tableau are a multiset with elements from a linearly ordered set, partitioned according to the rows of the tableau. A *shuffle* of the multiset is a permutation of the elements of the multiset such that each block is ordered by the linear order. For the tableau

$$\begin{pmatrix} a & a & \mathbf{c} & \mathbf{d} \\ \mathbf{b} & \mathbf{b} & \mathbf{c} & d \end{pmatrix},$$

the distinct multiset shuffles over the boldface letters are

$$\begin{pmatrix} a & a & c & d \\ b & b & c & d \end{pmatrix}, \begin{pmatrix} a & a & b & d \\ b & c & c & d \end{pmatrix}, \begin{pmatrix} a & a & b & c \\ b & c & d & d \end{pmatrix}, \begin{pmatrix} a & a & c & c \\ b & b & d & d \end{pmatrix}, \begin{pmatrix} a & a & b & b \\ c & c & d & d \end{pmatrix}.$$

The following theorem is the analogue for compact tableaux of the van der Waerden syzygy. We will use boldface to denote the set of letters to be shuffled in the identity.

THEOREM 4. *If T is a rectangular compact tableau with m rows, n columns, and at least $n+1$ letters in two rows boldface, and T has the property that if any letter is boldface, then all occurrences of that letter in the same row are boldface, then we have the following identity:*

$$\sum_{\delta}(c_{\delta_a}c_{\delta_b}c_{\delta_c}...c_{\delta_d})\ \delta\begin{pmatrix} & & \dots & X & \dots & & \\ \dots & A & \dots & \mathbf{a} & \dots & \mathbf{b} & \dots \\ \dots & \mathbf{c} & \dots & \mathbf{d} & \dots & E & \dots \\ & & \dots & Y & \dots & & \end{pmatrix} = 0,$$

where the sum ranges over all distinct multiset shuffles of the boldfaced letters and the coefficients, c_{δ_x}, are determined relative to the effect of δ on the boldface letters as follows:

for the letter x, δ moves $p-j$ x's into a row with j x's, $p \geq j \geq 0$.
1) if the j x's are boldface, then $c_{\delta_x} = 1$.
2) if the j x's are not boldface, then $c_{\delta_x} = \binom{p}{j}$.

This theorem, proved independently by McMillan (1990), also follows from the Exchange Lemma, p. 60, of Grosshans, Rota, and Stein (1987). See also Rota et al. (to appear).

EXAMPLE. The identity on the boldface letters of

$$\begin{pmatrix} a & b & \mathbf{c} & \mathbf{d} \\ \mathbf{b} & \mathbf{b} & \mathbf{c} & d \end{pmatrix}$$

gives

$$\begin{pmatrix} a & b & c & d \\ b & b & c & d \end{pmatrix} + 1\cdot 2\cdot 1\cdot 1\begin{pmatrix} a & b & b & d \\ b & c & c & d \end{pmatrix} + 1\cdot 2\cdot 1\cdot 2\begin{pmatrix} a & b & b & c \\ b & c & d & d \end{pmatrix}$$

$$+1\cdot 3\cdot 1\cdot 2\begin{pmatrix} a & b & b & b \\ c & c & d & d \end{pmatrix} + 1\cdot 1\cdot 1\cdot 2\begin{pmatrix} a & b & c & c \\ b & b & d & d \end{pmatrix} = 0.$$

Equivalently,

$$\begin{pmatrix} a & b & \mathbf{c} & \mathbf{d} \\ \mathbf{b} & \mathbf{b} & \mathbf{c} & d \end{pmatrix}$$

$$= -2\begin{pmatrix} a & b & b & d \\ b & c & c & d \end{pmatrix} - 4\begin{pmatrix} a & b & b & c \\ b & c & d & d \end{pmatrix} - 6\begin{pmatrix} a & b & b & b \\ c & c & d & d \end{pmatrix} - 2\begin{pmatrix} a & b & c & c \\ b & b & d & d. \end{pmatrix}$$

This is the form of the identity we will use, where we substitute the sum on the right side of the equation for the tableau on the left side. Note that in our notation, a tableau with boldface letters is equal to the same tableau without boldface, since the boldface notation is used only to inform the reader how Theorem 4 is being applied.

The entries of our tableaux are from a linearly ordered set. The signature convention allows us to order the entries of the rows as we wish. Now we adopt the convention of ordering the row entries in ascending order. We also order the rows lexicographically in ascending order, treating the rows as n-letter words. In the dotted bracket correspondence this is just a matter of commuting brackets in products and we record any sign changes imposed on our compact tableaux by the signature convention. With these ordering conventions, we can now impose an order on the tableaux. If T_1 and T_2 are $m \times n$ tableaux on the same letters and w_1 and w_2 are words obtained by concatenating the rows of the respective T_i, row 1 joined by row 2, etc., then we say $T_1 \leq T_2$ if and only if $w_1 \leq w_2$ (in the lexicographical order on the w_i). From now on when we refer to tableaux we will assume that this convention is adopted unless explicitly stated otherwise.

We define a *standard compact tableau* to be a tableau whose row entries are ascending and whose column entries are strictly ascending.

EXAMPLE.
$\begin{pmatrix} a & a & b & c \\ b & c & d & d \end{pmatrix}$ is a standard compact tableau, while

$\begin{pmatrix} a & a & b & d \\ b & c & c & d \end{pmatrix}$ is a compact tableau which is non-standard.

LEMMA 5. *Suppose T is a nonstandard compact tableau with entry y in row i and entry x in row j, with both entries in the same column and $x \leq y, i < j$. If we boldface y and all the entries of row i to the right of y together with all other occurrences of y in row i, and boldface x and all the entries of row j to the left of x together with all other occurrences of x in row j, and apply Theorem 4, T is a sum of compact tableaux $T = -\sum_\delta c_\delta T_\delta$ where $T_\delta < T$ for every δ.*

THEOREM 6 (THE DOTTED STRAIGHTENING ALGORITHM). *Every multilinear compact tableau expression is uniquely a linear combination of standard compact tableaux on the same multiset.*

EXAMPLE. In the examples, a tableau with boldface letters indicates that this tableau will be substituted for in the next step, using Lemma 5 on the boldface letters. We now completely straighten a non-standard tableau.

$$\begin{aligned}
\begin{pmatrix} a & a & \mathbf{c} & \mathbf{d} \\ \mathbf{b} & \mathbf{b} & \mathbf{c} & d \end{pmatrix} &= -\begin{pmatrix} a & a & b & \mathbf{d} \\ \mathbf{b} & \mathbf{c} & \mathbf{c} & \mathbf{d} \end{pmatrix} - 2\begin{pmatrix} a & a & b & c \\ b & c & d & d \end{pmatrix} - 2\begin{pmatrix} a & a & c & c \\ b & b & d & d \end{pmatrix} \\
&\quad - 2\begin{pmatrix} a & a & b & b \\ c & c & d & d \end{pmatrix} \\
&= 2\begin{pmatrix} a & a & b & b \\ c & c & d & d \end{pmatrix} + \begin{pmatrix} a & a & b & c \\ b & c & d & d \end{pmatrix} - 2\begin{pmatrix} a & a & b & c \\ b & c & d & d \end{pmatrix} \\
&\quad - 2\begin{pmatrix} a & a & c & c \\ b & b & d & d \end{pmatrix} - 2\begin{pmatrix} a & a & b & b \\ c & c & d & d \end{pmatrix} \\
&= -\begin{pmatrix} a & a & b & c \\ b & c & d & d \end{pmatrix} - 2\begin{pmatrix} a & a & c & c \\ b & b & d & d \end{pmatrix}.
\end{aligned}$$

It is easy to verify that a standard compact tableau is equal to a dotted bracket expression which, when expanded out as a signed sum of ordinary bracket monomials, is straightened in the usual sense. Thus, the ordinary straightening algorithm and the dotted straightening algorithm, when applied to a dotted bracket expression, accomplish the same calculation.

Both the ordinary and dotted straightening algorithms have time complexity which is quadratic in the total number of tableaux (modulo our ordering conventions) which are of the same shape as the input tableaux. It is clear that if we have a reasonable amount of dotting, there are considerably fewer compact tableaux than ordinary tableaux, and thus that the dotted straightening algorithm is much faster given a dotted bracket polynomial as input. The dotted straightening algorithm is particularly useful in the the Cayley factorization algorithm. Indeed, the first step of the factorization algorithm is to find the extensors, or sets of points in which the bracket polynomial is antisymmetric, and then to actually rewrite the polynomial so that it is explicitly dotted in those sets of points. Since this dotting has to be determined in any case, we do realize the efficiency improvement mentioned above.

References

Barnabei, M., Brini, A., Rota, G.-C. (1985). On the exterior calculus of invariant theory. *J. Algebra* **96**, 120-160.

Bokowski, J., Sturmfels, B. (1987). Polytopal and nonpolytopal spheres: an algorithmic approach. *Israel J. Math.* **57**, 257–271.

Chou, S., Schelter, W.F., Yang, J. (1987). Characteristic sets and Gröbner bases in geometry theorem proving. In:(H. Crapo, ed.) *Computer-aided geometric reasoning*, pp. 29–56. INRIA Rocquencourt, France.

Crapo, H. (1991). Invariant-theoretic methods in scene analysis and structural mechanics, *J. Symbolic Computation,* to appear.

Désarménien, J., Kung, J. P. S., Rota, G.-C. (1978). Invariant theory, Young bitableaux, and combinatorics. *Advances in Math.* **27**, 63-92.

Doubilet, P., Rota, G.-C., Stein, J. (1974). On the foundations of combinatorial theory IX: Combinatorial methods in invariant theory, *Studies in Applied Math.* **57**, 185–216.

Grosshans, F., Rota, G.-C., Stein, J. (1987). Invariant theory and superalgebras, C.B.M.S. Regional Conference Series, No. 69, *American Math. Society.*

Havel, T. (1991). The use of distances as coordinates in computer-aided proofs of theorems in Euclidean geometry, *J. Symbolic Computation,* to appear.

Kutzler, B., Stifter, S. (1986). On the application of Buchberger's algorithm to automated geometry theorem proving, *J. Symbolic Computation* **2**, 389-398.

McMillan, T. (1990). Invariants of Antisymmetric Tensors. Ph.D. thesis, University of Florida, Gainesville, FL, in preparation.

McMillan, T., White, N. (1991). The dotted straightening algorithm, *J. Symbolic Computation,* to appear.

Rota, G.-C., Stein, J. (1976). Applications of Cayley algebras, *Accademia Nazionale dei Lincei atti dei Convegni Lincei 17,* Colloquio Internazionale sulle Teorie Combinatoire, Tomo **2**, Roma.

Rota, G.-C., Stein, J. (1986). Symbolic method in invariant theory, *Proc. Nat. Acad. Sci. U.S.A., Mathematics* **83**, 844-847.

Rota, G.-C., Stein, J., Wang, R. (to appear). Supersymmetric algebra, supersymmetric spaces, and invariant theory.

Sturmfels, B., Whiteley, W. (1991). On the synthetic factorization of projectively invariant polynomials, *J. Symbolic Computation,* to appear.

White, N., (1991). Multilinear Cayley factorization, *J. Symbolic Computation,* to appear.

White, N., Whiteley, W. (1983). The algebraic geometry of stresses in frameworks, *S.I.A.M. J. Algebraic and Discrete Methods* **4**, 481-511.

White, N., Whiteley, W. (1987). The algebraic geometry of motions of bar-and-body frameworks, *S.I.A.M. J. Algebraic and Discrete Methods* **8**, 1-32.

Whiteley, W. (1990). The geometry of bivariate C_2^1-splines, preprint.

Young, A. (1928). On quantitative substitutional analysis, (3rd paper), *Proc. London Math Soc.*, Ser. 2, **28**, 255-292.

(Received: October 1990)

Acta Applicandae Mathematicae **21**: 185–192, 1990.

The Nagata-Higman Theorem

Edward Formanek

The Pennsylvania State University, University Park, PA 16802, U.S.A.

This article is devoted to a history of the Nagata-Higman Theorem and related developments. The presentation is for the most part in chronological order, with one important exception. Approximately five years ago, Gerald Schwarz discovered that the Nagata-Higman Theorem had been proved by Dubnov and Ivanov [2] in 1943, nine years before the paper of Nagata. It had been completely overlooked by the mathematical community and had had no impact on subsequent research. Therefore in my history it appears as a recent archaeological find.

The theorem proved by Nagata [5] in 1952 is

Theorem 1. *Let* A *be an algebra over a field* K *of characteristic zero. Suppose that there is a positive integer* n *such that* $a^n = 0$ *for all* $a \in A$. *Then there is an integer* N *such that* $a_1 a_2 \ldots a_N = 0$ *for all* $a_1, \ldots, a_N \in A$.

By an *algebra over* K we mean an associative ring, not necessarily with a unit, with an action of K which commutes with the ring multiplication. It is apparent that the theorem can be reformulated in terms of free algebras

AMS subject classification (1980). 15A72.
Key words. Rings of invariants, determinantal ideals.

and T-ideals. Let $K\langle X\rangle = K\langle x_1, x_2, \ldots\rangle$ be a free associative algebra without unit in countably many variables. A T-*ideal* is an ideal of $K\langle X\rangle$ which is closed under endomorphisms of $K\langle X\rangle$. Endomorphisms of $K\langle X\rangle$ are just specializations of the variables. The theorem then becomes

<u>Theorem</u> 2. *There is an integer* N *such that* $x_1 \ldots x_N$ *lies in the* T-*ideal of* $K\langle X\rangle$ *generated by* x_1^n.

Let $d(n)$ be the least integer such that $x_1 \ldots x_{d(n)}$ lies in the T-ideal $J(n)$ generated by x_1^n. Most subsequent research related to the Nagata-Higman Theorem has been concerned with the still unsolved problem of determining $d(n)$. A reasonable problem, which as far as I know has attracted no interest, is to find some concrete description of the relatively free ring $K\langle X\rangle/J(n)$. For example, let $V(m)$ be the K-vector space spanned by the $m!$ multilinear monomials in $x_1, \ldots, x_m$. Then $V(m)/(V(m) \cap J(n))$ is a module over S_n, the symmetric group on n letters, and one may try to determine the isomorphism type of this module. For arbitrary T-ideals, much work has been done in this direction, mainly by Regev. In the problem at hand, $d(n)$ is the least integer such that $V(d(n))/(V(d(n)) \cap J(n)) = 0$.

Nagata showed that the theorem is false if K is a field of characteristic p, where $p \leqq n$. He also obtained a very large upper bound

$f(n)$ for $d(n)$ and showed that the theorem holds in characteristic p, if $p > f(n)$. He conjectured that $p > n$ is sufficient, and the primary goal of Higman's 1956 paper [3] was proving this conjecture. However, Higman also gave a much simpler proof of the theorem, a much better upper bound for $d(n)$, namely 2^n-1 (or even $7\cdot 2^{n-3}-1$, for $n \geqq 3$), and the first lower bound for $d(n)$, namely n^2/e^2.

This lower bound was improved by Kuzmin [4] in 1975. He showed that $n(n+1)/2$ is a lower bound for $d(n)$, and conjectured that $d(n)$ is equal to $n(n+1)/2$. As far as I know, in published work his conjecture has been verified only for $n \leqq 3$.

The most interesting development since the original theorem came at this time from the work of Procesi and Razmyslov on the invariants of matrices. Their main result is that all relations between $n\times n$ matrices and their invariants are consequences of the Cayley-Hamilton polynomial. The fact that the leading term of the Cayley-Hamilton polynomial is x^n underlies the connection with the Nagata-Higman Theorem.

In order to describe their results more precisely, it is necessary to introduce two K-algebras associated with the invariants of matrices. Let $U(1), U(2), \ldots$ be $n\times n$ *generic matrices* over K. This means that $U(r) = (u_{ij}(r))$, $(1 \leqq i,j \leqq n,\ r = 1,2,\ldots)$, where the entries $u_{ij}(r)$ are independent commuting variables over K. The K-subalgebra of $K[u_{ij}(r)]$

generated by all traces $tr(\mu)$, where μ is a nontrivial monomial in the $U(r)$, is called the *ring of invariants of* $n\times n$ *matrices*, and is denoted $C(n)$. Let $R(n)$ be the $C(n)$-subalgebra (without unit) of $M_nK[u_{ij}(r)]$ generated by the generic matrices $U(r)$. As a $C(n)$-module, $R(n)$ is generated by all nontrivial monomials in the $U(r)$. Procesi showed [6, Corollary 4.7]

Theorem 3 . $K\langle X\rangle$ / {T-ideal generated by x_1^n } $\approx$ $R(n)$ / {ideal generated by $tr(\mu)$, μ a nontrivial monomial in the $U(r)$ }.

One interesting consequence of Theorem 3 is the fact that the T-ideal generated by x_1^n contains the T-ideal of identities of $n\times n$ matrices [6, Corollary 4.8]. It also leads to the following conclusion about $d(n)$, where $C(n)$ and $R(n)$ have the grading induced by letting each variable $u_{ij}(r)$ have degree one.

Theorem 4. (a) $d(n)$ *is the least integer such that* $R(n)$ *is generated as a* $C(n)$-*module by elements of degree* $\leqq d(n)-1$.

(b) $d(n)$ *is the least integer such that* $C(n)$ *is generated as a* K-*algebra by elements of degree* $\leqq d(n)$.

The second fundamental theorem of matrix invariants (Procesi [6, Theorem

4.3], Razmyslov [8, Proposition 1]) identifies the multilinear elements of $C(n)$ of degree r with a certain factor ring of KS_r, the group algebra of the symmetric group on r letters. This allows the determination of $d(n)$ to be reduced to a problem about the group algebra of the symmetric group. Recall that there is a one-to-one correspondence between Young diagrams (or partitions of r) and simple factors of KS_r. In this correspondence the two-sided ideal $I(n,r)$ corresponding to all partitions with more than n parts is equal to zero if $r \leqq n$, and is equal to the two-sided ideal generated by $\Sigma\ \{\ \mathrm{sign}(\pi)\pi \mid \pi \in S_{n+1}\ \}$ if $r \geqq n+1$.

<u>Theorem</u> 5. *Let* $Q(r+1)$ *be the* K-*vector subspace of* KS_{r+1} *spanned by all group elements except the* $(r+1)$-*cycles. Then* $d(n) \leqq r$ *if and only if* $KS_{r+1} = Q(r+1) + I(n,r+1)$.

Using the criterion of Theorem 5, Razmyslov [8, p. 759 of English translation] showed that $d(n) \leqq n^2$, which is still the best upper bound. His proof relied on the observation that a Young diagram with n^2+1 boxes either has $\geqq n+1$ boxes in the first row, or $\geqq n+1$ boxes in the first column. Recently, Strickland [9] translated Theorem 5 into a question involving determinantal ideals.

There has been no real progress on the determination of $d(n)$ since the work of Procesi and Razmyslov. But there was the very interesting rediscovery in 1985 by Gerald Schwarz of the work of Dubnov and Ivanov. In the above terminology, they showed that $R(n)$ is generated as a $C(n)$-algebra by monomials of degree $\leq 2^n-2$, and hence that the Nagata-Higman Theorem holds, with $d(n) \leq 2^n-1$. Even without knowing the isomorphism of Theorem 3, it is clear that a proof of the Nagata-Higman Theorem is implicit in their argument, for they manipulate the Cayley-Hamilton polynomial for $n\times n$ matrices

$$x^n + (\text{expression involving traces and lower powers of } x) = 0$$

in order to obtain a formula

$$x_1 \ldots x_N = (\text{expression involving traces and lower degree monomials}) ,$$

with $N = 2^n-1$. If the "traces" in their proof are thrown away, a proof of the Nagata-Higman Theorem very similar to Higman's is left. In addition, Dubnov [1] had shown in 1935 that $d(3) = 6$.

The reader may feel that, given the inequality $n(n+1)/2 \leq d(n) \leq n^2$, the exact value of $d(n)$ no longer is of sufficient interest. In that case he may find more stimulation in the 1988 theorem of Zelmanov [10], which settled affirmatively the important question of whether the n-th Engel condition implies Lie nilpotence in characteristic zero. Razmyslov [7] had constructed, for any prime $p > 5$, a nonnilpotent Lie algebra over a field of characteristic p which satisfies the $(p-2)$-th Engel condition. Here $[a,b] = ab-ba$, and higher commutators are defined inductively by $[a,b_1,\ldots b_{n+1}] = [[a,b_1,\ldots,b_n],b_{n+1}]$.

Theorem 6 (Zelmanov [10]). *Let* A *be an algebra over a field* K *of characteristic zero. Suppose that there is a positive integer* n *such that for all* $a,b \in A$, $[a,b,\ldots,b] = 0$, *where* n *is the number of* b*'s. Then there is an integer* N *such that* $[a,b_1,\ldots,b_N] = 0$ *for all* $a,b_1,\ldots,b_N \in A$.

As far as I know, no one has made a guess for the minimal value of N as a function of n. An upper bound could be obtained from Zelmanov's proof.

References

1. J. Dubnov, Proceedings of Seminar on Vector and Tensor Analysis, Mechanics Research Institute, Moscow State University (1935), 351-367 (Russian).

2. J. Dubnov and V. Ivanov, Sur l'abaissement du degre des polynomes en affineurs, C.R. (Doklady) Acad. Sci. URSS 41 (1943), 95-98.

3. G. Higman, On a conjecture of Nagata, Proc. Camb. Phil. Soc. 52 (1956), 1-4.

4. E. N. Kuzmin, On the Nagata-Higman Theorem, pp. 101-107 in *Mathematical structures - computational mathematics - mathematical modeling,* Proceedings dedicated to the sixtieth birthday of Academician L. Iliev, Sofia, 1975 (Russian).

5. M. Nagata, On the nilpotency of nil-algebras, J. Math. Soc. Japan 4 (1952), 296-301.

6. C. Procesi, The invariant theory of $n \times n$ matrices, Adv. in Math. 19 (1976), 306-381.

7. Y. P. Razmyslov, On Lie algebras satisfying the Engel condition, Alg. i Logika 10 (1971), 33-44 (Russian). English translation: Alg. and Logic 10 (1971), 21-29.

8. Y. P. Razmyslov, Trace identities of full matrix algebras over a field of characteristic zero, Izv. Akad. Nauk SSSR Ser. Mat. 38 (1974), 723-756 (Russian). English translation: Math. USSR Izv. 8 (1974), 727-760.

9. E. Strickland, On algebras satisfying the identity $X^n = 0$, Comm. Alg. 17 (1989), 2795-2799.

10. E. I. Zelmanov, Engelian Lie algebras, Sib. Mat. Zhur. 29 (1988), 112-117 (Russian). English translation: Sib. Math. Jour. 29 (1989), 777-781.

(Received: October 1990)

Acta Applicandae Mathematicae **21**: 193–246, 1990.

Supersymmetric Bracket Algebra and Invariant Theory

Rosa Q. Huang, Gian-Carlo Rota & Joel A. Stein

Department of Mathematics, MIT, Cambridge MA 02139, U.S.A.

AMS subject classifications (1980). 05A40, 15A72.
Key words. Supersymmetric algebra, matrix identities.

1 Supersymmetric Algebra

In this section we review some definitions and results in Ref. 9, pp 1-11. Roughly speaking, the supersymmetric algebra is a generalization of the ordinary algebra of polynomials in a set L of variables. Our variables shall be of two kinds: positively signed and negatively signed $L = L^+ \cup L^-$.

Let K always be an arbitrary infinite field, unless specially mentioned. Let L be a set of variables. To each $a \in L$ and $i \in \mathbf{N}$, we define a new variable $a^{(i)}$, called the ith **divided power** of a, with $a^{(0)} = 1$ and $a^{(1)} = a$. The **divided powers algebra** Divp$[L]$ is the ordinary polynomial algebra of the divided powers

$$\{a^{(i)} \mid a \in L, i \in \mathbf{N}\}$$

with the additional rule below, where $a^{(i)}$ behaves algebraically as $\frac{a^i}{i!}$:

$$a^{(i)}a^{(j)} = \binom{i+j}{i} a^{(i+j)}.$$

The divided powers algebra Divp$[L]$ seems similar to the ordinary polynomial algebra over the variable set L. But there is a subtle difference between them which makes the theory characteristic free.

The **exterior algebra** Ext$[L]$ over the variable set L is the algebra generated by the variables in L subject to the anticommutative rules:

(1) $ab = -ba$ for $a, b \in L$,

(2) $aa = 0$ for $a \in L$.

Now our variable set L will contain two kinds of variables, positively signed and negatively signed: $L = L^+ \cup L^-$. Such a set L of variables will be called a **proper signed set**. The **supersymmetric algebra** Super$[L]$ is the tensor product of the divided powers algebra Divp$[L^+]$ and the exterior algebra Ext$[L^-]$ with the rule that positively signed variables commute with negatively signed variables.

We proceed to give a more formal definition of Super$[L]$. Let Mon(L) be the free monoid generated by the variable set L, and let Div(L) be the free monoid generated by $L^- \cup \{a^{(i)} \mid a \in L, \ \ i \in \mathbf{N}\}$. Define the algebra Tens$[L]$ to be the free associative algebra generated by words in Div(L). For a word w in Div(L), define its **parity** $|w|$ to be 1 or 0 according to the number of occurrences of all negative variables in w is even or odd. The supersymmetric algebra Super$[L]$ is the quotient of Tens$[L]$ by the ideal I_L which is generated by all expressions of the following forms:

1.	$uv - (-1)^{\lvert u\rvert\lvert v\rvert}vu$ where $u, v \in \mathrm{Div}(L)$,
2.	aa where $a \in L^-$,
3.	$a^{(i)}a^{(j)} - \binom{i+j}{i}a^{(i+j)}$ where $a \in L^+$.

Let w be a word in Div(L). For a letter $a \in L$, the **content** of a in w, denoted by cont(w, a), is the relative degree of a in w with each divided power $a^{(i)}$ counted i-times. The **length** of w, written as length(w), is the total degree of w with each $a^{(i)}$ counted i-times. The **display** of w, denoted by disp(w), is a word in Mon(L) such that

- $\mathrm{disp}(a) = a$ for $a \in L^-$;
- $\mathrm{disp}(a^{(i)}) = aa \cdots a$ (i-times) for $a \in L^+$;
- $\mathrm{disp}(ww') = \mathrm{disp}(w)\mathrm{disp}(w')$ for $w, w' \in \mathrm{Div}(L)$.

The **support** of w, denoted by supp(w), is the subset of L such that

$$a \in \mathrm{supp}(w) \quad \text{if and only if} \quad \mathrm{cont}(w, a) \neq 0.$$

The **positive support** of w is defined by

$$\mathrm{supp}^+(w) = \mathrm{supp}(w) \cap L^+,$$

and the **negative support** of w is defined by

$$\text{supp}^-(w) = \text{supp}(w) \cap L^-.$$

Now let w be a word in Mon(L). The **stand** of w, denoted by stand(w), is the unique word in Div(L) such that the consecutive identical letters in w are contracted into one divided power.

Example. Let $w = a^{(2)}bcdaa \in \text{Div}(L)$, where $a, b \in L^+$ and $c, d \in L^-$. Then

$$|w| = 0, \quad \text{length}(w) = 7, \quad \text{cont}(w, a) = 4,$$

and

$$\text{disp}(w) = aabcdaa, \quad \text{stand}(aabcdaa) = a^{(2)}bcda^{(2)},$$

and

$$\text{supp}(w) = \{a, b, c, d\}, \quad \text{supp}^+(w) = \{a, b\}, \quad \text{supp}^-(w) = \{c, d\}.$$

2 The Coproduct Δ

Let Super[L]$\otimes$Super[L] be the tensor product of the algebra Super[L] with itself subject to the multiplicative rule

$$\boxed{u \otimes v \cdot u' \otimes v' = (-1)^{|v||u'|} uu' \otimes vv'.}$$

The supersymmetric algebra Super[L] has the structure of a Hopf algebra. The **c product** Δ which is an algebra homomorphism from Super[L] to Super[L]$\otimes$Super is defined such that:

1. $\Delta(1) = 1 \otimes 1,$
2. $\Delta(a) = 1 \otimes a + a \otimes 1$ if $a \in L^-,$
3. $\Delta(a^{(n)}) = a^{(n)} \otimes 1 + a^{(n-1)} \otimes a + \ldots + 1 \otimes a^{(n)}$ if $a \in L^+.$

If u is a monomial in Super$[L]$, then Δu is a sum of terms of the form $r \otimes s$, say

$$\Delta u = r_1 \otimes s_1 + r_2 \otimes s_2 + ... + r_p \otimes s_p,$$

we shall denote this sum by **Sweedler notation**,

$$\Delta u = \sum_u u_{(1)} \otimes u_{(2)}.$$

For example, let $a \in L^+$ and let $b, c \in L^-$, then

$$\Delta(a^{(2)}b) = a^{(2)}b \otimes 1 + ab \otimes a + a^{(2)} \otimes b + b \otimes a^{(2)} + a \otimes ab + 1 \otimes a^{(2)}b,$$

and

$$\Delta(bc) = bc \otimes 1 + b \otimes c - c \otimes b + 1 \otimes bc.$$

The **coassociative law**

$$(1 \otimes \Delta) \cdot \Delta = (\Delta \otimes 1) \cdot \Delta$$

holds for the coproduct Δ, where each side of the identity is an algebra homomorphism from Super$[L]$ to Super$[L] \otimes$ Super$[L] \otimes$ Super$[L]$. (The detail can be found in Ref. 9.) Therefore by the recursive relations

$$\Delta^{(k)} = (1 \otimes \Delta) \cdot \Delta^{(k-1)}$$

we can define a sequence of algebra homomorphisms

$$\Delta^{(k)} : \text{Super}[L] \to \text{Super}[L] \otimes ... \otimes \text{Super}[L] \quad (\ (k+1)\text{-}times\ ).$$

The image of a monomial u in Super$[L]$ will be written as

$$\Delta^{(k)}u = \sum_u u_{(1)} \otimes u_{(2)} \otimes ... \otimes u_{(k)}.$$

The **augmentation** map ε : Super$[L] \to K$ is defined as follows. For a monomial u, let

$$\varepsilon(u) = 1 \ \text{ if length}(u) = 0, \qquad \varepsilon(u) = 0 \ \text{ if length}(u) > 0,$$

then extend ε linearly to an algebra homomorphism. The **antipode** s of the Hopf algebra Super[L] is the linear extension of the map

$$s(u) = (-1)^k u,$$

where u is a monomial of length k. The antipode s and the coproduct Δ are related by the identity

$$\boxed{\sum_u s(u_{(1)})u_{(2)} = \sum_u u_{(1)}s(u_{(2)}) = \varepsilon(u)}$$

where u is a monomial in Super[L].

3 Polarizations

In this section we review some materials in Ref. 18, especially the concepts of polarization operators $D(b,a)$ and word polarization operators $T(w,a)$. Roughly speaking, polarizations $D(b,a)$ are signed derivations in the supersymmetric algebra Super[L] which replace a letter a by a letter b. When the letters a and b are of the same sign, the polarization $D(b,a)$ is a positive derivation; when the letters a and b are of different signs, the polarization $D(b,a)$ is a negative derivation. A rigorous definition is the follows.

A **polarization operator** $D(b,a)$ of Super[L] is a linear operator defined recursively by the rules below, where w and w' are arbitrary monomials in Super[L]:

1. If $D(b,a)$ is a negative polarization, i.e., the letters a and b are of different signs, then we set
   - $D(b,a)1 = 0$,
   - $D(b,a)c = 0 \quad \text{if} \quad c \neq a$,

- $D(b,a)a^{(k)} = ba^{(k-1)}$ if $a \in L^+$,
- $D(b,a)a = b$ if $a \in L^-$,
- $D(b,a)(ww') = D(b,a)w \cdot w' + (-1)^{|w|} w \cdot D(b,a)w'$.

2. If $D(b,a)$ is a positive polarization, i.e. the letters a and b are of the same sign, then we set

   - $D(b,a)1 = 0$,
   - $D(b,a)c = 0$ if $c \neq a$,
   - $D(b,a)a^{(k)} = ba^{(k-1)}$ if $a \in L^+$,
   - $D(b,a)a = b$ if $a \in L^-$,
   - $D(b,a)(ww') = D(b,a)w \cdot w' + w \cdot D(b,a)w'$.

3. If $D(b,a)$ is a positive polarization, then we set the divided powers

   - $D^{(0)}(b,a)w = w$,
   - $D^{(1)}(b,a) = D(b,a)$,
   - $D^{(k)}(b,a)a^{(r)} = 0$ if $r < k$,
   - $D^{(k)}(b,a)a^{(r)} = b^{(k)}a^{(r-k)}$ if $r \geq k$,
   - $D^{(k)}(b,a)(ww') = \sum_{i=0}^{k} D^{(i)}(b,a)w \cdot D^{(k-i)}(b,a)w'$.

4. If $D(b,a)$ is a negative polarization we set

   - $D^{(0)}(b,a)w = w$,
   - $D^{(1)}(b,a) = D(b,a)$,
   - $D^{(k)}(b,a) = 0$ if $k \geq 1$.

Proposition 1 *The polarization operators $D(b,a)$ of Super[L] are well defined.*

Proof. Define polarization operators $\mathbf{D}(\mathbf{b},\mathbf{a})$ and their divided powers $\mathbf{D}^{(\mathbf{k})}(\mathbf{b},\mathbf{a})$ on the algebra Tens$[L]$ by the same rules above. We have to check that the ideal I_L is closed under these operators. In view of the rules

$$\mathbf{D}(\mathbf{b},\mathbf{a})(ww') = \mathbf{D}(\mathbf{b},\mathbf{a})w \cdot w' + (-1)^{|w|} w \cdot \mathbf{D}(\mathbf{b},\mathbf{a})w'$$

when $\mathbf{D}(\mathbf{b},\mathbf{a})$ is a negative polarization, and

$$\mathbf{D}^{(\mathbf{k})}(\mathbf{b},\mathbf{a})(ww') = \sum_{i=0}^{k} \mathbf{D}^{(\mathbf{i})}(\mathbf{b},\mathbf{a})w \cdot \mathbf{D}^{(\mathbf{k-i})}(\mathbf{b},\mathbf{a})w'$$

when $\mathbf{D}(\mathbf{b},\mathbf{a})$ is a positive polarization, it is sufficient to check for all polarizations $\mathbf{D}^{(\mathbf{k})}(\mathbf{b},\mathbf{a})$ that

- $\mathbf{D}^{(\mathbf{k})}(\mathbf{b},\mathbf{a})(uv - (-1)^{|u||v|}vu) \in I_L \quad \text{for} \quad u, v \in \mathrm{Div}(L)$,
- $\mathbf{D}^{(\mathbf{k})}(\mathbf{b},\mathbf{a})cc \in I_L \quad \text{for} \quad c \in L^-$,
- $\mathbf{D}^{(\mathbf{k})}(\mathbf{b},\mathbf{a})(c^{(i)}c^{(j)} - \binom{i+j}{i}c^{(i+j)}) \in I_L \quad \text{for} \quad c \in L^+$.

Most of the checking is routine. Here we only check the first one when $\mathbf{D}(\mathbf{b},\mathbf{a})$ is a positive polarization. Notice that in this case

$$|\mathbf{D}^{(\mathbf{k})}(\mathbf{b},\mathbf{a})u| = |u|.$$

So we get

$$\begin{aligned}
&\mathbf{D}^{(\mathbf{k})}(\mathbf{b},\mathbf{a})(uv - (-1)^{|u||v|}vu) \\
= &\sum_{i=0}^{k} \mathbf{D}^{(\mathbf{i})}(\mathbf{b},\mathbf{a})u \cdot \mathbf{D}^{(\mathbf{k-i})}(\mathbf{b},\mathbf{a})v - (-1)^{|u||v|} \sum_{i=0}^{k} \mathbf{D}^{(\mathbf{k-i})}(\mathbf{b},\mathbf{a})v \cdot \mathbf{D}^{(\mathbf{i})}(\mathbf{b},\mathbf{a})u \\
= &\sum_{i=0}^{k} (\mathbf{D}^{(\mathbf{i})}(\mathbf{b},\mathbf{a})u \cdot \mathbf{D}^{(\mathbf{k-i})}(\mathbf{b},\mathbf{a})v \\
&-(-1)^{|\mathbf{D}^{(\mathbf{k-i})}(\mathbf{b},\mathbf{a})v||\mathbf{D}^{(\mathbf{i})}(\mathbf{b},\mathbf{a})u|} \mathbf{D}^{(\mathbf{k-i})}(\mathbf{b},\mathbf{a})v \cdot \mathbf{D}^{(\mathbf{i})}(\mathbf{b},\mathbf{a})u),
\end{aligned}$$

which is an element in I_L as desired. ∎

Positive polarizations are abstractions from the classical differential operators

$$\sum_{i=1}^{n} b_i \frac{\partial}{\partial a_i},$$

whereas negative polarizations have the property that $D(b,a)^2 = 0$. They are boundary operators in the sense of homological algebra, and can be used to define complexes.

Let w be a word in $\mathrm{Div}(L)$, and let a be any positive letter. The **word polarization** operator $T(w,a)$, or simply $T(w)$ when no confusion occurs, is uniquely defined by the following requirements:

1. $T(b,a) = D(b,a) \quad \text{if} \quad b \in L,$
2. $T(b^{(k)},a) = D^{(k)}(b,a) \quad \text{if} \quad b \in L^+,$
3. $T(ww',a) = T(w,a)T(w',a).$

When $\mathrm{supp}(u) \cap \mathrm{supp}(w) = \emptyset$, there are nice properties about $T(w)u$.

Proposition 2 *Let w be a word of length k, and let a be a positive letter such that $a \notin supp(w)$, then*

$$T(w,a)a^{(k)} = w.$$

Proof. Let $w = b^{(i)}c^{(j)} \cdots d^{(l)}$, then

$$\begin{aligned} & T(w,a)a^{(k)} \\ = \;& T(b^{(i)}c^{(j)} \cdots d^{(l)}, a)a^{(k)} \\ = \;& T(b^{(i)},a)\, T(c^{(j)},a) \;\cdots\; T(d^{(l)},a)\; a^{(k)} \\ = \;& b^{(i)}c^{(j)} \cdots d^{(l)} \\ = \;& w \end{aligned}$$

which completes the proof. ∎

Proposition 3 *Let L_1 and L_2 be two disjoint proper signed sets, and let a be a positive letter in L_2. Then $T(u,a)w$ can be expanded to a bilinear operator from Super$[L_1] \times$ Super$[L_2]$ to Super$[L_1 \cup L_2]$ which sends the pair (u,w) to $T(u,a)w$, with $T(1,a)w = w$.*

Proof. Let L be a proper signed set. Notice that

$$T(uv,a) = T(u,a)\,T(v,a),$$

it is sufficient to prove the following:

- $T(uv - (-1)^{|u||v|}vu, a)w = 0$,
  where $u,v,w \in \mathrm{Div}(L)$, $\mathrm{supp}(uv) \cap \mathrm{supp}(w) = \emptyset$,
- $T(bb,a)w = 0$ for $b \in L^-$,
- $T(b^{(i)}b^{(j)} - \binom{i+j}{i} b^{(i+j)}, a)w = 0$ for $b \in L^+$.

Then the proof follows obviously if we write $w = a^{(k)}w'$ where $a \notin \mathrm{supp}(w')$. ■

Proposition 4 *Let u and w be two monomials in Super[L] such that supp$(u) \cap$ supp$(w) = \emptyset$, and let a be a positive letter in supp(w). Then*

$$\boxed{\Delta T(w,a)u = \textstyle\sum_{u,w}(-1)^{|u_{(1)}||w_{(2)}|}T(w_{(1)},a)u_{(1)} \otimes T(w_{(2)},a)u_{(2)}.}$$

Proof. Let L_1 and L_2 be two disjoint proper signed sets. For a monomial $w \in \mathrm{Super}[L_1]$ and a positive letter $a \in L_2$, define a bilinear map

$$\mathbf{T(w,a)}: \ \mathrm{Super}[L_2] \times \mathrm{Super}[L_2] \to \mathrm{Super}[L_2] \otimes \mathrm{Super}[L_2]$$

such that

$$\mathbf{T(w,a)}(u,v) = \sum_w (-1)^{|u||w_{(2)}|}T(w_{(1)},a)u \otimes T(w_{(2)},a)v.$$

e well-definedness of $\mathbf{T(w,a)}$ is an immediate consequence of the well-definedness the operator $T(w,a)$ on Super$[L]$. Therefore $\mathbf{T(w,a)}$ induces a map

$$T(w,a): \text{ Super}[L_2] \otimes \text{Super}[L_2] \to \text{Super}[L_2] \otimes \text{Super}[L_2]$$

ch that

$$T(w,a)u \otimes v = \sum_w (-1)^{|u||w_{(2)}|} T(w_{(1)},a)u \otimes T(w_{(2)},a)v.$$

it is sufficient to prove that

$$T(w,a) \circ \Delta = \Delta \circ T(w,a). \quad (1)$$

fact it is enough to prove that

$$D^{(k)}(b,a) \circ \Delta = \Delta \circ D^{(k)}(b,a), \quad (2)$$

cause if $w = w'b^{(k)}$, then

$$T(w,a) \circ \Delta = T(w',a) \circ D^{(k)}(b,a) \circ \Delta = T(w',a) \circ \Delta \circ D^{(k)}(b,a),$$

erefore the proof of identity (1) can be carried out by induction on length(w). ext we proceed to prove identity (2). Let $a^{(i)}u$ be a monomial in Super$[L_2]$ here $a \notin \text{supp}(u)$ and $i \geq k$, then

$$\begin{aligned} & D^{(k)}(b,a) \circ \Delta(a^{(i)}u) \\ = \; & D^{(k)}(b,a)\Big(\sum_u \sum_{s+t=i} a^{(s)}u_{(1)} \otimes a^{(t)}u_{(2)}\Big) \\ = \; & \sum_u \sum_{s+t=i} \sum_{p+q=k} (-1)^{q|b||u_{(1)}|} a^{(s-p)}b^{(p)}u_{(1)} \otimes a^{(t-q)}b^{(q)}u_{(2)} \end{aligned}$$

here we put $a^{(j)} = b^{(j)} = 0$ if $j < 0$. The last expression equals

$$\Delta(a^{(i-k)}b^{(k)}u) = \Delta(D^{(k)}(b,a)(a^{(i)}u)).$$

herefore (2) is proved. ∎

4 The Algebra of Braces

with L a proper signed set, let {L} be the variable set whose members are all monomials w in Super$[L]$. A letter in {L} will be denoted by $\{w\}$. The **parity** $|\{w\}|$ of the letter $\{w\}$ is defined to be the parity of the integer $|w| + \text{length}(w)$. We denote by Brace$\{L\}$ the associative algebra over the field K generated by the alphabet $\{L\}$, subject to the following commutative relations:

- $\{w\}\{w'\} = (-1)^{(|\{w\}||\{w'\}|)}\{w'\}\{w\}$,
- $\{w\}\{w\} = 0 \quad \text{if} \quad |\{w\}| = 1$.

If $p, q, r \in$ Super$[L]$, and if $\alpha \in K$, then the element $\{p\}$ of Brace$\{L\}$ is defined recursively by the rules

- $\{p\} = \{q\} + \{r\} \quad \text{if} \quad p = q + r$,
- $\{\alpha p\} = \alpha\{p\}$.

If $D(b, a)$ is a polarization operator of Super$[L]$, the linear operator $\mathbf{D(b, a)}$ of Brace$\{L\}$ is defined recursively by the rules below, where m and m' are arbitrary monomials in Brace$\{L\}$:

1. For all polarizations $D(b, a)$ and for all $p \in$ Super$[L]$, we set

$$\mathbf{D(b, a)}\{p\} = \{D(b, a)p\}.$$

2. If $D(b, a)$ is a negative polarization, we set

   - $\mathbf{D^{(k)}(b, a)}m = 0 \quad \text{for} \quad k > 1$,
   - $\mathbf{D(b, a)}(mm') = \mathbf{D(b, a)}m \cdot m' + (-1)^{|m|}m \cdot \mathbf{D(b, a)}m'$,

   where the parity of a monomial $m = \{w\}\{w'\}...\{w''\}$ in Brace$\{L\}$ is the sum

$$|m| = |\{w\}| + |\{w'\}| + ... + |\{w''\}|.$$

3. If $D(b, a)$ is a positive polarization, we set

- $\mathbf{D^{(k)}(b,a)}\{w\} = \{D^{(k)}(b,a)w\}$,
- $\mathbf{D^{(k)}(b,a)}(mm') = \sum_{i=0}^{k} \mathbf{D^{(i)}(b,a)}m \cdot \mathbf{D^{(k-i)}(b,a)}m'$.

Similarly we can define word polarizations $\mathbf{T(w,a)}$ from $T(w,a)$ in the obvious way.

For example, if a and b are negative letters and if $\text{cont}(u,a) = \text{cont}(u',a) = 0$ for some monomials $u, u' \in \text{Super}[L]$, we have

$$\mathbf{D^{(2)}(b,a)}(\{au\}\{au'\}) = \{bu\}\{bu'\}.$$

An element of the algebra of braces is called **brace polynomial.** An element of the brace algebra which is a product of braces is called a **brace monomial.** A brace polynomial is said to be **homogeneous** if it is the sum of monomials in each of which every letter occurs with the same content, but different letters may occur with different contents.

To avoid unnecessary minus signs, it turns out to be convenient to introduce the **vertical notation** for monomials. The monomial $\text{vert}\{w, w', ..., w''\}$ is defined as

$$\boxed{\text{vert}\{w, w', ..., w''\} = \mathbf{T(w,a)T(w',b)}\cdots\mathbf{T(w'',c)}(\{a^{(i)}\}\{b^{(j)}\}\cdots\{c^{(k)}\})}$$

where $\text{cont}(w,a) = ... = \text{cont}(w'',c) = 0$, and $\text{length}(w) = i$, ..., $\text{length}(w'') = k$. We may also write the above monomial in the **tableau notation**:

$$\left\{\begin{array}{c} w \\ w' \\ \vdots \\ w'' \end{array}\right\} = \text{vert}\{w, w', ..., w''\},$$

this monomial differs at most by a sign from the monomial $\{w\}\{w'\}...\{w''\}$. For example,

$$\text{vert}\{w\} = \{w\}, \qquad \text{and}$$

$$\text{vert}\{w, w'\} = \left\{ \begin{array}{c} w \\ w' \end{array} \right\} = (-1)^{k|w'|}\{w\}\{w'\},$$

where $k = \text{length}(w)$.

In vertical notation we have

$$\mathbf{D}^{(\mathbf{k})}(\mathbf{b}, \mathbf{a})\text{vert}\{w, w'\} = \sum_{i=0}^{k}(-1)^{(k-i)|w|}\text{vert}\{\mathbf{D}^{(\mathbf{i})}(\mathbf{b}, \mathbf{a})w, \mathbf{D}^{(\mathbf{k-i})}(\mathbf{b}, \mathbf{a})w'\}$$

when $\mathbf{D}(\mathbf{b}, \mathbf{a})$ is a negative polarization, and

$$\mathbf{D}^{(\mathbf{k})}(\mathbf{b}, \mathbf{a})\text{vert}\{w, w'\} = \sum_{i=0}^{k}\text{vert}\{\mathbf{D}^{(\mathbf{i})}(\mathbf{b}, \mathbf{a})w, \mathbf{D}^{(\mathbf{k-i})}(\mathbf{b}, \mathbf{a})w'\}$$

when $\mathbf{D}(\mathbf{b}, \mathbf{a})$ is a positive polarization.

5 The Bracket Algebra

With L a proper signed alphabet, we proceed to define an algebraic system that connects the notion of a supersymmetric algebra with a symbolic representation of projective invariants. This is the algebra of brackets over the supersymmetric algebra Super$[L]$.

The **bracket algebra** Bracket$[L]$ of **dimension** n over the field K (dependence on the field K and on the dimension n will not be indicated) is the quotient of the algebra Brace$\{L\}$ obtained by imposing the congruence relations below, where $w, w', ..., w''$ are any monomials in Super$[L]$:

(1) $\{w\} = 0 \quad \text{if} \quad \text{length}(w) \neq n.$

(2) $\{w\}\{w'\}...\{w''\} = 0$ whenever any positive letter a of L occurs more than n times in the monomial $\{w\}\{w'\}...\{w''\}$, i.e.,

$$\text{cont}(\{w\}\{w'\}...\{w''\}, a) = \text{cont}(w, a) + ... + \text{cont}(w'', a) > n.$$

(3) Let $\{w\}\{w'\}...\{w''\}$ be a monomial in Brace$\{L\}$ in which some positive letter a occurs more than n times, and let $b, c, d, e, ..., f, g$ be any letters in L. Then

$$\mathbf{D^{(i)}(b,c)D^{(j)}(d,e)\cdots D^{(k)}(f,g)}(\{w\}\{w'\}...\{w''\}) = 0.$$

The canonical image of an element p of Brace$\{L\}$ in Bracket$[L]$ will be denoted by Image(p). In particular, we write Image$(\{w\}) = [w]$, and

$$\text{Image}(\{w\}\{w'\}...\{w''\}) = [w][w']\cdots[w'].$$

Proposition 5 *Every polarization operator* $\mathbf{D^{(k)}(b,a)}$ *on Brace*$\{L\}$ *induces a polarization operator on Bracket[L].*

This is an immediate consequence of the condition (3) above.

We will again denote by $D^{(k)}(b,a)$ the induced polarizations $\mathbf{D^{(k)}(b,a)}$. We write $|[w]| = |\{w\}|$ when w is a word of length n.

Proposition 6 *Let p be an element of Brace*$\{L\}$ *such that Image*$(p) = 0$. *Then Image*$(\mathbf{D^{(k)}(b,a)}p) = 0$ *for any polarization* $\mathbf{D(b,a)}$ *and for any positive integer* k. *More generally, for any brace polynomial p, we have*

$$Image(\mathbf{D^{(k)}(b,a)}p) = D^{(k)}(b,a)Image(p).$$

Proof. It is sufficient to prove the identity for a brace monomial

$$p = \{w_1\}\{w_2\}\cdots\{w_m\}$$

where w_i are monomials of length n in Super$[L]$. we have

$$\begin{aligned}
&\text{Image}(D^{(k)}(b,a)(\{w_1\}\{w_2\}\cdots\{w_m\})) \\
= \;& \text{Image}(\sum_{k_1+\cdots k_m=k} \varepsilon(k_1,k_2,...,k_m)\{D^{(k_1)}(b,a)w_1\}\cdots\{D^{(k_m)}(b,a)w_m\}) \\
= \;& \sum_{k_1+\cdots k_m=k} \varepsilon(k_1,k_2,...,k_m)[D^{(k_1)}(b,a)w_1]\cdots[D^{(k_m)}(b,a)w_m] \\
= \;& D^{(k)}(b,a)([w_1][w_2]\cdots[w_m]) \\
= \;& D^{(k)}(b,a)(\text{Image}(D^{(k)}(b,a)(\{w_1\}\{w_2\}\cdots\{w_m\}))
\end{aligned}$$

where

$$\varepsilon(k_1, k_2, ..., k_m) = 1$$

if $D(b, a)$ is a positive polarization, and

$$\varepsilon(k_1, k_2, ..., k_m) = (-1)^{k_2|[w_1]|+k_3|[w_1][w_2]|+\cdots+k_m|[w_1]\cdots[w_{m-1}]|}$$

if $D(b, a)$ is a negative polarization. ∎

An element p of the bracket algebra Bracket$[L]$ will be called a **bracket polynomial.** A product of brackets will be called a **bracket monomial. Homogeneous** bracket polynomials are defined as the images of homogeneous brace polynomials. The **vertical notation** and **tableau notation** for bracket monomials are similarly defined as

$$\begin{bmatrix} w \\ w' \\ \vdots \\ w'' \end{bmatrix} = \text{vert}[w, w', ..., w''] = \text{Image}(\text{vert}\{w, w', , , , , w''\}),$$

that is

$$\text{vert}[w, w', ..., w''] = T(w, a)T(w', b)...T(w'', c)([a^{(n)}][b^{(n)}]...[c^{(n)}]),$$

where $a, b, ..., c$ are not in supp(w), supp(w'), ..., supp(w'') respectively, and where the word polarization $T(w, a)$ of Bracket$[L]$ is defined in the obvious way from polarizations $D(b, a)$.

6 The Exchange Identity

We proceed to establish the main results holding for the algebra of brackets.

Proposition 7 *Let w be a monomial of length greater than n in Super[L], and let w' be any monomial in Super[L]. Then*

$$\sum_w \pm[w_{(1)}][w_{(2)}w'] = \sum_w vert[w_{(1)}, w_{(2)}w'] = 0. \tag{3}$$

Note that the vertical notation for monomials gives an automatic rule for the sign on the far left side.

Proof. Suppose length$(w) = p > n$. Choose a positive letter $a \notin \text{supp}(w)$ and a positive letter $b \notin \text{supp}(w')$. Then the left side of (3) is the image in Bracket$[L]$ of the brace polynomial

$$\mathbf{T}(\mathbf{w},\mathbf{a})\mathbf{T}(\mathbf{w}',\mathbf{b})\{a^{(n)}\}\{a^{(p-n)}b^{2n-p)}\} = \sum_w \text{vert}\{w_{(1)}, w_{(2)}w'\},$$

and the image of the above brace polynomial is evidently 0, since it is a polarization of a brace monomial containing more than n occurrences of the positive letter a. ∎

Before we state the Exchange Identity, which is the main identity holding for Bracket$[L]$, we shall give the following identity in the Hopf algebra Super$[L]$, which can be found in Ref. 18.

Lemma 1 *Let u, v, w be arbitrary monomials in Super[L], then*

$$\sum_v uv_{(1)} \otimes v_{(2)} = (-1)^{|u||v|} \sum_u \sum_{vu_{(1)}} (vu_{(1)})_{(1)} \otimes (vu_{(1)})_{(2)} s(u_{(2)}).$$

Proof. From the identity

$$\sum_w w_{(1)} \cdot s(w_{(2)}) = \varepsilon(w),$$

we have

$$\begin{aligned}
&\sum_u \sum_{vu_{(1)}} (vu_{(1)})_{(1)} \otimes (vu_{(1)})_{(2)} s(u_{(2)}) \\
= \; &\sum_u \sum_v (-1)^{|u_{(11)}||v_{(2)}|} v_{(1)} u_{(11)} \otimes v_{(2)} u_{(12)} s(u_{(2)}) \\
= \; &\sum_v \sum_u \sum_{u_{(2)}} (-1)^{|u_{(1)}||v_{(2)}|} v_{(1)} u_{(1)} \otimes v_{(2)} u_{(21)} s(u_{(22)}) \\
= \; &\sum_v (-1)^{|u||v_{(2)}|} v_{(1)} u \otimes v_{(2)} \\
= \; &(-1)^{|u||v|} \sum_v uv_{(1)} \otimes v_{(2)},
\end{aligned}$$

which completes the proof. ∎

Proposition 8 (**Exchange Identity**) *Let u, v, w be monomials in Super[L]. Then the following identity holds:*

$$\sum_v \pm[uv_{(1)}][v_{(2)}w] = \sum_u \pm[vu_{(1)}][s(u_{(2)})w].$$

More precisely, the signs are determined as follows:

$$\boxed{\sum_v vert[uv_{(1)}, v_{(2)}w] = (-1)^{|u||v|}\sum_u vert[vu_{(1)}, s(u_{(2)})w].}$$

Proof. Starting with the identity

$$\sum_v uv_{(1)} \otimes v_{(2)} = (-1)^{|u||v|}\sum_u \sum_{vu_{(1)}} (vu_{(1)})_{(1)} \otimes (vu_{(1)})_{(2)}s(u_{(2)}),$$

we obtain from the universal property of tensor products that

$$\sum_v \text{vert}\{uv_{(1)}, v_{(2)}w\} = (-1)^{|u||v|}\sum_u \sum_{vu_{(1)}} \text{vert}\{(vu_{(1)})_{(1)}, (vu_{(1)})_{(2)}s(u_{(2)})w\}.$$

by the proceeding proposition. If length$(vu_{(1)}) > n$, the image in Bracket$[L]$ of the corresponding term on the right side equals 0, and therefore the right side is congruent to

$$(-1)^{|u||v|}\sum_u \text{vert}\{vu_{(1)}, s(u_{(2)})w\},$$

as desired. ∎

When only positive letters are involved in u, v, w, the Exchange Identity becomes

$$\boxed{\sum_v [v_{(1)}u][v_{(2)}w] = \sum_u [u_{(1)}v][s(u_{(2)})w].}$$

Proposition 9 *The quotient of the brace algebra Brace$\{L\}$ by the congruence relations generated by the following identities:*

(1) $\{w\} = 0$ *if length*$(w) \neq n$,

(2) the Exchange Identity

$$\sum_v vert\{uv_{(1)}, v_{(2)}w\} = (-1)^{|u||v|}\sum_u vert\{vu_{(1)}, s(u_{(2)})w\},$$

is naturally isomorphic to the bracket algebra Bracket[L].

Proof. We first show that if m is a brace monomial containing more than n occurrences of a positive letter a, then m is congruent to 0. Suppose that

$$m = \text{vert}\{a^{(k)}w, a^{(r)}w'\}, \quad \text{where} \quad k+r>n.$$

The Exchange Identity gives

$$\text{vert}\{a^{(k)}w, a^{(r)}w'\} = \sum_w \text{vert}\{a^{(k+r)}w_{(1)}, s(w_{(2)})w'\}$$

which is congruent to 0, since $k+r>n$. If the bracket monomial m contains more than two factors, one extends the proceeding argument by induction.

Next we prove invariance under polarizations. If p is congruent to 0 in the quotient algebra defined above, then we show that

$$\mathbf{D^{(k)}(b,a)}p$$

is congruent to 0 in the quotient algebra. This will follow from noting that polarization in Super[L] preserves the degree of a monomial in Super[L] and that polarization commutes with the coproduct (see Proposition 4) and the antipode map, i.e.,

$$\begin{aligned} & D^{(k)}(b,a)\sum_w w_{(1)}\otimes w_{(2)} \\ = & \sum_{i,w} D^{(i)}(b,a)w_{(1)}\otimes D^{(k-i)}(b,a)w_{(2)} \\ = & \sum_{D^{(k)}(b,a)w} (D^{(k)}(b,a)w)_{(1)}\otimes (D^{(k)}(b,a)w)_{(2)}, \end{aligned}$$

and

$$D^{(k)}(b,a)s(w) = s(D^{(k)}(b,a)w).$$

In view of the identity satisfied by polarizations in Brace$\{L\}$

$$\mathbf{D^{(k)}(b,a)}(pq) = \sum_{i=0}^{k} \pm \mathbf{D^{(i)}(b,a)}p\cdot \mathbf{D^{(k-i)}(b,a)}q,$$

we need only show that the defining Exchange Identity in this proposition is preserved under polarization. This we proceed to verify.

First note that polarizations preserve the length of a brace. Therefore it is enough to verify that under polarization, Exchange Identities are mapped into linear combinations of Exchange Identities. We have two cases and we will use the tableau notation in the following discussions.

Case 1. $\mathbf{D}(\mathbf{b}, \mathbf{a})$ is a positive polarization.

Consider the expression

$$\mathbf{D}^{(\mathbf{k})}(\mathbf{b}, \mathbf{a}) \left(\sum_v \left\{ \begin{array}{cc} u & v_{(1)} \\ v_{(2)} & w \end{array} \right\} - (-1)^{|u||v|} \sum_u \left\{ \begin{array}{cc} v & u_{(1)} \\ s(u_{(2)}) & w \end{array} \right\} \right).$$

Expand this expression using the rules given previously for polarizing expressions of the form vert$\{w, w'\}$ in Brace$\{L\}$ and expressions of the form mm' in Super$[L]$. The expression which results is

$$\sum_{i,j,r,t} \sum_v \left\{ \begin{array}{cc} D^{(i)}(b,a)u & D^{(j)}(b,a)v_{(1)} \\ D^{(r)}(b,a)v_{(2)} & D^{(t)}(b,a)w \end{array} \right\}$$
$$-(-1)^{|u||v|} \sum_{i,j,r,t} \sum_u \left\{ \begin{array}{cc} D^{(i)}(b,a)v & D^{(j)}(b,a)u_{(1)} \\ D^{(r)}(b,a)s(u_{(2)}) & D^{(t)}(b,a)w \end{array} \right\},$$

where the outmost sums range over all non-negative integers i, j, r, t which sum to k. Since polarization commutes with the antipode map, we infer that the above expressions equals

$$\sum_{i,j,r,t} \sum_v \left\{ \begin{array}{cc} D^{(i)}(b,a)u & D^{(j)}(b,a)v_{(1)} \\ D^{(r)}(b,a)v_{(2)} & D^{(t)}(b,a)w \end{array} \right\}$$
$$-(-1)^{|u||v|} \sum_{i,j,r,t} \sum_u \left\{ \begin{array}{cc} D^{(i)}(b,a)v & D^{(j)}(b,a)u_{(1)} \\ s(D^{(r)}(b,a)u_{(2)}) & D^{(t)}(b,a)w \end{array} \right\}.$$

Since polarization commutes with the coproduct Δ, the latest expression is seen to equal

$$\sum_{i,j,t}\sum_{v}\left\{\begin{array}{cc} D^{(i)}(b,a)u & (D^{(j)}(b,a)v)_{(1)} \\ (D^{(j)}(b,a)v)_{(2)} & D^{(t)}(b,a)w \end{array}\right\}$$
$$-(-1)^{|u||v|}\sum_{i,j,t}\sum_{u}\left\{\begin{array}{cc} D^{(i)}(b,a)v & (D^{(j)}(b,a)u)_{(1)} \\ s((D^{(j)}(b,a)u)_{(2)}) & D^{(t)}(b,a)w \end{array}\right\},$$

where the outmost sums now range over all non-negative integers i, j, t which sum to k. Since $\mathbf{D}(\mathbf{b},\mathbf{a})$ is a positive polarization, we have that whenever $D^{(j)}(b,a)u$ and $D^{(i)}(b,a)v$ are non-zero,

$$|u| = |D^{(i)}(b,a)u| \quad \text{and} \quad |v| = |D^{(i)}(b,a)v|.$$

The above sum therefore equals

$$\sum_{i,j,t}\sum_{v}\left\{\begin{array}{cc} D^{(i)}(b,a)u & (D^{(j)}(b,a)v)_{(1)} \\ (D^{(j)}(b,a)v)_{(2)} & D^{(t)}(b,a)w \end{array}\right\}$$
$$-\sum_{i,j,t}(-1)^{|D^{(j)}(b,a)u||D^{(i)}(b,a)v|}\sum_{u}\left\{\begin{array}{cc} D^{(i)}(b,a)v & (D^{(j)}(b,a)u)_{(1)} \\ s((D^{(j)}(b,a)u)_{(2)}) & D^{(t)}(b,a)w \end{array}\right\}.$$

Interchanging i and j in the first innermost sum, the above expression is seen to equal

$$\sum_{i,j,t}\sum_{v}\left\{\begin{array}{cc} D^{(j)}(b,a)u & (D^{(i)}(b,a)v)_{(1)} \\ (D^{(i)}(b,a)v)_{(2)} & D^{(t)}(b,a)w \end{array}\right\}$$
$$-\sum_{i,j,t}(-1)^{|D^{(j)}(b,a)u||D^{(i)}(b,a)v|}\sum_{u}\left\{\begin{array}{cc} D^{(i)}(b,a)v & (D^{(j)}(b,a)u)_{(1)} \\ s((D^{(j)}(b,a)u)_{(2)}) & D^{(t)}(b,a)w \end{array}\right\},$$

which is a linear combination of the Exchange Identities as desired.

Case 2. $\mathbf{D}(\mathbf{b},\mathbf{a})$ is a negative polarization.

Reasoning as in case 1 we have:

$$\mathbf{D}(\mathbf{b},\mathbf{a})\left(\sum_{v}\left\{\begin{array}{cc} u & v_{(1)} \\ v_{(2)} & w \end{array}\right\} - (-1)^{|u||v|}\sum_{u}\left\{\begin{array}{cc} v & u_{(1)} \\ s(u_{(2)}) & w \end{array}\right\}\right)$$

$$
\begin{aligned}
&= \sum_{i,j,r,t} \sum_{v} (-1)^{j|u|+r|u|+r|v_{(1)}|+t|u|+t|v|} \left\{ \begin{array}{cc} D^{(i)}(b,a)u & D^{(j)}(b,a)v_{(1)} \\ D^{(r)}(b,a)v_{(2)} & D^{(t)}(b,a)w \end{array} \right\} \\
&- \ (-1)^{|u||v|} \sum_{i,j,r,t} \sum_{u} (-1)^{j|v|+r|v|+r|u_{(1)}|+t|u|+t|v|} \left\{ \begin{array}{cc} D^{(i)}(b,a)v & D^{(j)}(b,a)u_{(1)} \\ D^{(r)}(b,a)s(u_{(2)}) & D^{(t)}(b,a)w \end{array} \right.
\end{aligned}
$$

where the outmost sums range over all non-negative integers i, j, r, t which sum to 1 (one). Since polarization commutes with the antipode map, this expression equals in turn

$$
\begin{aligned}
& \sum_{i,j,r,t} \sum_{v} (-1)^{j|u|+r|u|+r|v_{(1)}|+t|u|+t|v|} \left\{ \begin{array}{cc} D^{(i)}(b,a)u & D^{(j)}(b,a)v_{(1)} \\ D^{(r)}(b,a)v_{(2)} & D^{(t)}(b,a)w \end{array} \right\} \\
-(-1)^{|u||v|} \quad & \sum_{i,j,r,t} \sum_{u} (-1)^{j|v|+r|v|+r|u_{(1)}|+t|u|+t|v|} \left\{ \begin{array}{cc} D^{(i)}(b,a)v & D^{(j)}(b,a)u_{(1)} \\ s(D^{(r)}(b,a)u_{(2)}) & D^{(t)}(b,a)w \end{array} \right.
\end{aligned}
$$

and since polarization commutes with the coproduct Δ, the last-given expression equals

$$
\begin{aligned}
& \sum_{i,j,t} \sum_{v} (-1)^{(j|u|+t|u|+t|v|} \left\{ \begin{array}{cc} D^{(i)}(b,a)u & (D^{(j)}(b,a)v)_{(1)} \\ (D^{(j)}(b,a)v)_{(2)} & D^{(t)}(b,a)w \end{array} \right\} \\
-(-1)^{|u||v|} \quad & \sum_{i,j,t} \sum_{u} (-1)^{j|v|+t|u|+t|v|} \left\{ \begin{array}{cc} D^{(i)}(b,a)v & (D^{(j)}(b,a)u)_{(1)} \\ s((D^{(j)}(b,a)u)_{(2)}) & D^{(t)}(b,a)w \end{array} \right\},
\end{aligned}
$$

where the outmost sums now range over all non-negative integers i, j, t which sum to 1. Interchanging i and j in the first innermost sum, we see that the above expression equals

$$
\begin{aligned}
& \sum_{i,j,t} \sum_{v} (-1)^{i|u|+t|u|+t|v|} \left\{ \begin{array}{cc} D^{(j)}(b,a)u & (D^{(i)}(b,a)v)_{(1)} \\ (D^{(i)}(b,a)v)_{(2)} & D^{(t)}(b,a)w \end{array} \right\} \\
-(-1)^{|u||v|} \quad & \sum_{i,j,t} \sum_{u} (-1)^{j|v|+t|u|+t|v|} \left\{ \begin{array}{cc} D^{(i)}(b,a)v & (D^{(j)}(b,a)u)_{(1)} \\ s((D^{(j)}(b,a)u)_{(2)}) & D^{(t)}(b,a)w \end{array} \right\},
\end{aligned}
$$

Noting that when $D^{(i)}(b,a)v$ and $D^{(j)}(b,a)u$ are non-zero, we have

$$
(-1)^{|D^{(i)}(b,a)v||D^{(j)}(b,a)u|} = (-1)^{(i+|v|)(j+|u|)}
$$

and that $ij = 0$, we see that the above expression equals

$$\sum_{i,j,t}\sum_{v}(-1)^{i|u|+t|u|+t|v|}\left\{\begin{array}{cc} D^{(j)}(b,a)u & (D^{(i)}(b,a)v)_{(1)} \\ (D^{(i)}(b,a)v)_{(2)} & D^{(t)}(b,a)w \end{array}\right\}$$

$$\sum_{i,j,t}(-1)^{|D^{(j)}(b,a)u||D^{(i)}(b,a)v|}\sum_{D^{(j)}(b,a)u}(-1)^{i|u|+t|u|+t|v|}\left\{\begin{array}{cc} D^{(i)}(b,a)v & (D^{(j)}(b,a)u)_{(1)} \\ s((D^{(j)}(b,a)u)_{(2)}) & D^{(t)}(b,a)w \end{array}\right\},$$

which equals in turn

$$\sum_{i,j,t}(-1)^{i|u|+t|u|+t|v|}\sum_{v}\left\{\begin{array}{cc} D^{(j)}(b,a)u & (D^{(i)}(b,a)v)_{(1)} \\ (D^{(i)}(b,a)v)_{(2)} & D^{(t)}(b,a)w \end{array}\right\}$$

$$\sum_{i,j,t}(-1)^{i|u|+t|u|+t|v|}\cdot(-1)^{|D^{(j)}(b,a)u||D^{(i)}(b,a)v|}\sum_{D^{(j)}(b,a)u}\left\{\begin{array}{cc} D^{(i)}(b,a)v & (D^{(j)}(b,a)u)_{(1)} \\ s((D^{(j)}(b,a)u)_{(2)}) & D^{(t)}(b,a)w \end{array}\right\},$$

the above expression is a linear combination of the Exchange Identities as desired.

The proof is complete. ■

7 Standard Basis Theorem

In this section we derive a straightening formula for the bracket algebra Bracket$[L]$, which is a special case of the straightening formula for double tableaux described in Ref. 9, in which we take $P = P^- = \{1, 2, ..., n\}$. This will be one of the main combinatorial tools to be applied in invariant theory.

Let L be a proper signed set. A **Young diagram** on L, $D = (w_1, w_2, ..., w_k)$, is a sequence of words $w_i \in \mathrm{Mon}(A)$ such that $\lambda_1 \geq \lambda_2 \geq ... \geq \lambda_k$ where $\lambda_i = \mathrm{length}(w_i)$. The vector $\lambda = (\lambda_1, \lambda_2, ..., \lambda_k)$ is called the **shape** of D. The word w_i is called the ith row of the Young diagram D. The integer k is called the **height** of D, denoted by height(D). We define **content** of this setting by

$$\mathrm{cont}(D) = \mathrm{cont}(w_1) + \mathrm{cont}(w_2) + ... + \mathrm{cont}(w_k)$$

If $w_1 = x_1x_2...x_{\lambda_1}, w_2 = y_1y_2...y_{\lambda_2}, ...$, then we envision D as

$$\begin{array}{c} x_1x_2...x_{\lambda 1} \\ y_1y_2...y_{\lambda 2} \\ \vdots \end{array}$$

The word $\tilde{w}_i = x_iy_i...$ is called the ith column of D.

A linearly ordered proper signed set L is called an **alphabet.** From now on let L be an alphabet and let $D = (w_1, w_2, ..., w_k)$ be a Young diagram on L. We call D **standard** when for each of the words

$$\begin{aligned} w_i &= x_1x_2...x_r, \quad x_j \in L, \\ \tilde{w}_i &= y_1y_2...y_s, \quad y_j \in L, \end{aligned}$$

the following conditions are satisfied:

(1) if $x_j \in L^+$, then $x_j \leq x_{j+1}$,
(2) if $x_j \in L^-$, then $x_j < x_{j+1}$,
(3) if $y_j \in L^+$, then $y_j < y_{j+1}$,
(4) if $y_j \in L^-$, then $y_j \leq y_{j+1}$.

The **Young tableau** of D is the monomial in Bracket$[L]$

$$\text{Young}(D) = [\text{stand}(w_1)][\text{stand}(w_2)]...[\text{stand}(w_k)].$$

The Young tableau Young(D) does not vanish only if length$(w_i) = n$ for each row w_i. Later we will denote by $[D]$ the Young tableau Young(D) when no confusion occurs.

Theorem 1 *Let L be an alphabet. The $[D_i]$, where D_i are standard Young diagrams, form a basis of Bracket[L] over K.*
In particular, let D be a Young diagram, then

$$\boxed{[D] = \textstyle\sum_i c_i[D_i]}$$

where D_i have the same height and content as D, and where the integer coefficients c_i are unique.

We now proceed to prove Theorem 1. The spanning part is almost same as the proof in Ref. 9 where it deals with a more general case, double tableaux. The uniqueness of the spanning is proved by using a fact about Capelli operator in Ref. 4 and Ref. 5.

Proof of the spanning.

Let $D = (w_1, w_2, ..., w_k)$. We prove that $[D]$ is a linear combination of $[D_i]$ with integer coefficients , where D_i are standard Young diagrams.

If $k = 1$, it is obvious, since we can arrange the letters in w_1 in a non-decreasing order.

Next, suppose $k > 1$. We shall carry out the proof in a subspace of Bracket$[L]$ over K spanned by all Young tableaux $[D]$ of the given content.

We now suppose that the spanning part of this theorem is false for some Young diagrams D. We order such diagrams via the lexicographic order obtained by lining up the rows of D. That is , suppose $w_i = x_{i1}x_{i2}...x_{i\lambda_i}$, we apply the lexicographic order to the word

$$x_{11}x_{12}...x_{1\lambda_1}x_{21}x_{22}...x_{k\lambda_k}.$$

Among the diagrams for which the spanning part of the theorem is false, we choose the smallest D in this lexicographic order. It follows that the rows of D are non-decreasing, i.e.,

$$x_{i1} \leq x_{i2} \leq ... \leq x_{i\lambda_i}.$$

If D has a row of repeated negative letter, then $[D] = 0$ and we are done. Otherwise, since D is not standard, conditions (3) or (4) in the definition of standard diagrams must be violated. We shall assume that the first such violation occurs in rows

$$x_1x_2...x_mx_{m+1}x_{m+2}...x_n$$

$$y_1y_2\ldots y_m y_{m+1} y_{m+2}\ldots y_n$$

when comparing x_{m+1} to y_{m+1}.

Case 1. $x_{m+1} > y_{m+1}$. We begin by noting that

$$x_1 \le \ldots \le x_m < x_{m+1} \le \ldots \le x_n,$$

$$y_1 \le \ldots \le y_m \le y_{m+1} = \ldots = y_{m+k} < y_{m+k+1} \le \ldots \le y_n.$$

Therefore

$$y_1 \le y_2 \le \cdots \le y_{m+k} < x_{m+1} \le \cdots \le y_n.$$

In the Exchange Identity

$$\sum_v \begin{bmatrix} u & v_{(1)} \\ v_{(2)} & w \end{bmatrix} = (-1)^{|u||v|} \sum_u \begin{bmatrix} v & u_{(1)} \\ s(u_{(2)}) & w \end{bmatrix} \tag{4}$$

we take

$$u = \text{stand}(x_1\ldots x_m), \quad v = \text{stand}(x_{m+1}\ldots x_n y_1\ldots y_{m+k}), \quad w = \text{stand}(y_{m+k+1}\ldots y_n).$$

Since length$(v) > n$, (4) becomes

$$\sum_v \begin{bmatrix} u & v_{(1)} \\ v_{(2)} & w \end{bmatrix} = 0. \tag{5}$$

Now we examine the terms on the left-hand side. The term $[D]$ appears once. In the other terms $v_{(1)}$ must contain letters from $y_1y_2\ldots y_{m+k}$; the corresponding Young diagram is smaller than D in the lexicographic order. So the spanning part of the Theorem holds for D (which is a contradiction).

Case 2. $x_{m+1} = y_{m+1} = z$, which is a positive letter in L. Suppose that the rows in the question are

$$x_1\ldots x_m \overbrace{z\ldots z}^{i} x_{m+i+1}\ldots x_n$$

and

$$y_1 \dots y_k \underbrace{z \dots z}^{j} y_{k+j+1} \dots y_n,$$

where

$$x_1 \leq \dots \leq x_m < z < x_{m+i+1} \leq \dots \leq x_n$$

and

$$y_1 \leq \dots \leq y_k < z < y_{k+j+1} \leq \dots \leq y_n.$$

Again, we use the Exchange Identity (4), but now with

$$u = \text{stand}(x_1 \dots x_m), \quad v = z^{(i+j)}\text{stand}(x_{m+i+1} \dots x_n y_1 \dots y_k), \quad w = \text{stand}(y_{k+j+1} \dots y_n).$$

Since length$(v) > n$, we still get

$$\sum_v \begin{bmatrix} u & v_{(1)} \\ v_{(2)} & w \end{bmatrix} = 0.$$

One of the terms on the left-hand side is $[D]$. In the other terms $v_{(1)}$ either contains more of the letter z or contains letters from $y_1 y_2 \dots y_k$; the corresponding Young diagram is smaller in the lexicographic order. The proof of the spanning part is complete.

Proof of the Uniqueness.

We show that Young tableaux $[D_i]$ are linearly independent if D_i are standard. We may assume that D_i have the given content and the alphabet L is large enough to carry the proof.

Suppose there is a non-trivial linear relation

$$\sum_i c_i[D_i] = 0 \quad \text{where} \quad c_i \in K. \tag{6}$$

We first prove that if there are any positive letters occurring in D_i, then we can get a non-trivial linear relation among $[E_i]$, where E_i are standard Young diagrams on negative letters L^- of L, by polarizing away the positive letters. Suppose a positive letter a occurs in D_i with cont$(D_i, a) = k > 0$, we choose k different negative letters $b_1, b_2, \dots, b_k$ such that

(1) $b_j \notin \text{supp}(D_i)$ for all b_j,

(2) $b_1 < b_2 < ... < b_k$,

(3) $x < a < y$ if and only if $x < b_j < y$ for all $x, y \in \text{supp}(D_i)$ and for all b_j.

Let $T = D(b_1, a)D(b_2, a)...D(b_k, a)$, then for each D_i, we have a sum

$$T([D_i]) = [D_{i1}] + [D_{i2}] + ... + [D_{ik_i}]$$

where D_{ij} is the set of all different Young diagrams obtained from D_i by replacing each a by a letter b_l. (Assume that the letters $b_1, b_2, ..., b_k$ are arranged in an increasing order within each row of D_{ij}.) We observe that

(1) D_{ij} are standard diagrams;

(2) D_{ij} and $D_{i'j'}$ are different diagrams if $D_i \neq D_{i'}$.

Therefore apply the operator T to both sides of (6), we get a non-trivial linear relation

$$\sum_i c_i T[D_i] = \sum_i c_i([D_{i1}] + [D_{i2}] + ... + [D_{ik_i}]) = 0.$$

If there are still some positive letters occurring in D_{ij}, we proceed the same process above until all the positive letters are gone.

Next we restrict that all letters of L are negative and show that it is impossible to have a non-trivial linear relation among $[D_i]$, where D_i are standard diagrams of the same height and content.

For a Young diagram D, we define the Capelli operator $C(D)$ on Bracket$[L]$ as in Ref. 4:

$$C(D) = \prod_{x \in L} \prod_{i=1}^{n} D^{(\alpha(i,x))}(y_i, x)$$

where $\alpha(i, x)$ is the number of occurrences of the letter x in the ith column of the diagram D, and where $y_1, y_2, ..., y_n$ are n different letters of L and none of y_i belongs to supp(D).

Let $\tilde{w}_1, \tilde{w}_2, ..., \tilde{w}_n$ be the n columns of a Young diagram D. We order diagrams D of the given content by the lexicographic order of the word $\tilde{w}_1\tilde{w}_2...\tilde{w}_n$ in $\text{Mon}(L)$.

Let D_1 and D_2 be two standard Young diagrams of the same content, suppose D_2 is larger than D_1 in the lexicographic order of the associated column sequences, then the Capelli operator $C(D_1)$ has the properties (see Ref. 4 for details):

1. $C(D_1)([D_1]) \neq 0,$
2. $C(D_1)([D_2]) = 0.$

Suppose we have a linear relation

$$c_1[D_1] + c_2[D_2] + ... + c_k[D_k] = 0$$

where $c_1 \neq 0$, and where $D_1, D_2, ..., D_k$ are standard diagrams of the same content and $D_1 < D_2 < ... < D_k$ in the lexicographic order. We apply the operator $C(D_1)$ to both sides of the above equation, it becomes

$$c_1 C(D_1)([D_1]) = 0.$$

hence $c_1 = 0$, which is a contradiction. The proof is complete. ■

8 Models: the Classical Invariant Theory

Let $L = L^-$ be a negative alphabet (in this section only).

Then the bracket algebra Bracket$[L]$ is commutative. Our objective is to define the notion of model, a word borrowed from logic, for the bracket algebra. The concept of model was first introduced in Ref. 10.

Our models will be Peano spaces in the sense defined in Ref. 1, which we now recall. We define a **bracket** (of step n) on the vector space V of dimension

n over the field K to be a non-degenerate n-linear K-valued form defined over the vector space V, in symbols, a function

$$x_1, x_2, ..., x_n \longrightarrow [x_1, x_2, ..., x_n]$$

defined as the vectors $x_1, x_2, ..., x_n$ range over the vector space V, with the following properties:

(1) $[x_1, x_2, ..., x_n] = 0$ if at least two among the x's coincide. This implies that the bracket is a skew-symmetric multilinear form;

(2) for every x, y in V and every choice of scalars α, β in K, we have

$$\begin{aligned} & [x_1, x_2, ..., x_{i-1}, \alpha x + \beta y, x_{i+1}, ..., x_n] \\ = \; & \alpha[x_1, x_2, ..., x_{i-1}, x, x_{i+1}, ..., x_n] + \beta[x_1, x_2, ..., x_{i-1}, y, x_{i+1}, ..., x_n]. \end{aligned}$$

A sequence of vectors $e_1, e_2, ..., e_n$, necessarily a basis of V, such that $[e_1, e_2, ..., e_n]$ 1, will be called a **unimodular** basis. We define a **Peano space** of step n to be a pair $(V,[\,])$, where v is a vector space of dimension n and $[\,]$ is a bracket of step n over V. We shall denote a Peano space by the single letter V, leaving the bracket understood, whenever no confusion is possible.

A peano space can be viewed geometrically as a vector space in which an oriented volume element is specified. The bracket $[x_1, x_2, ..., x_n]$ gives the signed volume of the paralleelepiped spanned by the vectors x_i. The $[x_1, x_2, ..., x_n] = 0$ will hold if and only if the vectors are linearly dependent.

Every bracket can be expressed as a determinant. Indeed, choose any unimodular basis $e_1, e_2, ..., e_n$. Let $x_{ij} = [e_1, e_2, ..., e_{j-1}, x_i, e_{j+1}, ..., e_n]$. Then it is easily verified that $[x_1, x_2, ..., x_n] = \det(x_{ij})$. Such an expression for the bracket as a determinant is misleading. The bracket should be visualized as an object existing in its own right, independently of the choice of a coordinate system, much like an inner product defining a Hilbert space is to be visualized in its own right.

A **model** of the ring Bracket$[L]$, with its value in a Peano space V, is a function ϕ from the alphabet L to the Peano space V. The **induced map** Φ on brackets defined as

$$\boxed{\Phi([a_1 a_2 ... a_n]) = [\phi(a_1), \phi(a_2), ..., \phi(a_n)]}$$

extends to a homomorphism Φ of the bracket algebra Bracket$[L]$ to the field K, which will called the **induced homomorphism** of the model ϕ. Note that the bracket on the right is the bracket of the Peano space V, whereas the bracket on the left is the bracket of the bracket algebra Bracket$[L]$.

The fact that the map Φ is indeed a homomorphism is a consequence of the fact that the bracket in the Peano space satisfies the Exchange Identity. Since the bracket in a peano space is a determinant, the Exchange Identity can be derived from the classical Laplace expansions of a determinant. (see Ref. 1 for details.)

A model ϕ having the property that there exist letters $f_1, f_2, ..., f_n$ in L such that

$$\Phi([f_1 f_2 ... f_n]) = 1$$

is said to be a **unimodular model** relative to $f_1, f_2, \cdots, f_n$. There is no loss of generality in assuming that a model is unimodular, and we shall do so from now on.

For any letter a in L other than $f_1, f_2, ..., f_n$, consider n variables, labeled a_j, where $1 \leq j \leq n$. Let Pol(L) be the polynomial algebra over the field K of the variables set a_j, where $1 \leq j \leq n$ and where a ranges over L. We have

Proposition 10 *The map*

$$[f_1 f_2 ... f_{j-1} a f_{j+1} ... f_n] \to a_j$$

extends to an isomorphism of the quotient of Bracket[L] by the equation $[f_1 f_2 ... f_n]$ 1 *onto the polynomial algebra Pol(L) which maps* $[a_1 a_2 ... a_n]$ *to* $det(a_{ij})$.

Proof. Denote Bracket$[L]'$ as the quotient algebra of Bracket$[L]$ by the equation $[f_1 f_2 \cdots f_n] = 1$. Let

$$\tilde{L} = L \setminus \{f_1, f_2, \cdots, f_n\}.$$

Given a bracket $[w]$ in Bracket$[L]'$, if w contains more than one letters in $\tilde{L}$, we write

$$w = au,$$

then by the Exchange Identity ,

$$\begin{aligned} [w] &= [au] = [au][f_1 f_2 \cdots f_n] \\ &= [f_1 u][a f_2 \cdots f_n] - [f_2 u][f_1 a f_3 \cdots f_n] + \cdots + (-1)^{n-1}[f_n u][f_1 \cdots f_{n-1} a]. \end{aligned}$$

Therefore if $[w]$ contains k letters in $\tilde{L}$, then $[w][f_1 f_2 \cdots f_n]^{k-1}$ can be written as a polynomial in

$$\left\{[f_1 \cdots f_{j-1} a f_{j+1} \cdots f_n] \mid a \in \tilde{L},\ j = 1, 2, \cdots, n\right\}.$$

Hence we may assume that any element in Bracket$[L]'$ is such a polynomial. Then it is obvious to see that the map

$$\psi : \text{Pol}(L) \to \text{Bracket}[L]'$$

such that

$$a_j \to [f_1 \cdots f_{j-1} a f_{j+1} \cdots f_n]$$

is the inverse of the map defined in the proposition. The rest of the proof follows easily. ∎

Proposition 11 *Let ϕ be a unimodular model of Bracket[L] relative to $f_1, f_2, \cdots, f_n$ into the Peano space V. Then there is a unique map $\tilde{\phi}$ of the polynomial algebra Pol(L) to the field K such that $\tilde{\phi}(a_j)$ is the jth coordinate of the vector $\phi(a)$ in the coordinate system $\phi(f_1), \phi(f_2), ..., \phi(f_n)$ in V.*
Indeed, we have

$$\begin{aligned}\tilde{\phi}(a_j) &= \Phi([f_1 f_2 ... f_{j-1} a f_{j+1} ... f_n]) \\ &= [\phi(f_1), \phi(f_2), ..., \phi(f_{j-1}), \phi(a), \phi(f_{j+1}), \cdots, \phi(f_n)].\end{aligned}$$

This is a corollary of the previous proposition.

The following result states that every identity satisfied by determinants is a consequence of the Exchange Identity.

Proposition 12 *(Second fundamental theorem of invariant theory, classical version.) Let p be a polynomial in Bracket[L] such that $\Phi(p) = 0$ for every model ϕ. Then $p = 0$.*

In other words, any invariantly defined expression $p = 0$ can be expressed by an equation $q = 0$, where q is a bracket polynomial not containing the coordinates.

Proof. Let $p(a, b, \cdots, c)$ be a bracket polynomial in Bracket$[L]$ such that $\Phi(p) = 0$ for all models ϕ. Without loss of generality we assume that p is a sum of monomials which have the same content. So the notation $\text{cont}(p, a)$ makes sense.
For each $x \in L$, $\lambda \in K$, we have

$$p(a + \lambda x, b, \cdots, c) = \sum_{i=0}^{k} \lambda^i D^{(i)}(x, a) p(a, b, \cdots, c) \tag{7}$$

where $k = \text{cont}(p, a)$. For each model ϕ, we also have

$$\Phi(p(a + \lambda x, b, \cdots, c)) = 0.$$

Since K is an infinite field, we infer from (7) that for each $1 \leq i \leq k$,

$$\Phi(D^{(i)}(x,a)p(a,b,\cdots,c)) = 0 \tag{8}$$

for all models ϕ.

Let $x_1, x_2, \cdots, x_k$ be k different letters which does not occur in p, then in the bracket polynomial

$$\begin{aligned} & T(x_1x_2\cdots x_k,\ a)p(a,b,\cdots,c) \\ = \ & D(x_1,a)D(x_2,a)\cdots D(x_k,a)p(a,b,\cdots,c), \end{aligned}$$

k of the letter a are replaced by shuffles of k different letters $x_1, x_2, \cdots, x_k$. Therefore we may assume that each letter a occurs in p at most once.

If p does not vanish in Bracket[L], by the Standard Basis Theorem,

$$p = \sum_{i=1}^{k} c_i[D_i], \qquad c_i \neq 0, \tag{9}$$

where D_i are standard diagrams. Suppose

$$D_1 < D_2 < \cdots < D_k$$

in the lexicographic order of the sequence obtained by lining up each columns of D_i. Since each letter occurs at most once in D_1, the Capelli operator $C(D_1)$ becomes

$$C(D_1) = \prod_{i+1}^{n} \prod_{a} D(x_i, a)$$

where the inner product ranges over the letters a occurring in the ith column of D_1.

Apply the Capelli operator to both sides of (9), getting

$$C(D_1)p = c_1 C(D_1) D_1 = c_1 [x_1 x_2 \cdots x_n]^t$$

where t is the number of rows in D_1. Therefore by (8) we have

$$\Phi([x_1 x_2 \cdots x_n]^t) = 0$$

for all models ϕ, which obviously a contradiction, since we may choose a model ϕ such that $\phi(x_1), \phi(x_2), \cdots, \phi(x_n)$ form a basis of V.
Therefore p has to vanish in Bracket$[L]$ if $\Phi(p) = 0$ for all models ϕ. ■

Let $p(a, b, ..., c; f_1, f_2, ..., f_n)$ be a polynomial in the brackets

$$[f_1 f_2 ... f_{j-1} a f_{j+1} ... f_n], \ [f_1 f_2 ... f_{j-1} b f_{j+1} ... f_n], \ ..., \ [f_1 f_2 ... f_{j-1} c f_{j+1} ... f_n].$$

which is homogeneous in the letters $a, b, \cdots, c$. We say that p is an **invariant** of the letters $a, b, \cdots, c$ when , if ϕ and ϕ' are unimodular models relative to $f_1, f_2, \cdots, f_n$ such that

$$\phi(a) = \phi'(a), \ \phi(b) = \phi'(b), \ ..., \ \phi(c) = \phi'(c) \tag{10}$$

we have

$$\Phi(p) = \Phi'(p). \tag{11}$$

Given two models ϕ and ϕ' satisfied (10), suppose for some set of n negative letters $e_1, e_2, \cdots, e_n \in L$

$$\phi(e_i) = \phi'(f_i), \qquad i = 1, 2, \cdots, n, \tag{12}$$

then (11) becomes

$$\Phi(p(a, b, \cdots, c; f_1, f_2, \cdots, f_n)) = \Phi(p(a, b, \cdots, c; e_1, e_2, \cdots, e_n)). \tag{13}$$

Since there is no loss of generality to assume (12), the polynomial p is an invariant if and only if (13) holds for all models ϕ such that

$$[\phi(f_1), \phi(f_2), \cdots, \phi(f_n)] = [\phi(e_1), \phi(e_2), \cdots, \phi(e_n)] = 1.$$

Proposition 13 *(First fundamental theorem of invariant theory, classical version). Let p be an invariant of $a, b, ..., c$, then there exists a bracket polynomial q, not containing $f_1, f_2, ..., f_n$, such that*

$$\boxed{p(a, b, \cdots, c; f_1, f_2, \cdots, f_n) = [f_1 f_2 ... f_n]^g q(a, b, \cdots, c)}$$

for some non-negative integer g.

Proof. (The idea is due to G.-C. Rota.) For an arbitrary model ϕ satisfying

$$[\phi(f_1), \phi(f_2), \cdots, \phi(f_n)] \neq 0 \quad \text{and} \quad [\phi(e_1), \phi(e_2), \cdots, \phi(e_n)] \neq 0,$$

define a new model $\tilde{\phi}$ such that

$$\tilde{\phi}(x) = \phi(x), \quad \text{for} \quad x \in L,\ x \neq f_1,\ x \neq e_1,$$

and

$$\begin{aligned} \tilde{\phi}(f_1) &= [\phi(f_1), \phi(f_2), \cdots, \phi(f_n)]^{-1}\phi(f_1), \\ \tilde{\phi}(e_1) &= [\phi(e_1), \phi(e_2), \cdots, \phi(e_n)]^{-1}\phi(e_1), \end{aligned}$$

then

$$[\tilde{\phi}(f_1), \tilde{\phi}(f_2), \cdots, \tilde{\phi}(f_n)] = [\tilde{\phi}(e_1), \tilde{\phi}(e_2), \cdots, \tilde{\phi}(e_n)] = 1,$$

therefore the condition (13) is true for $\tilde{\phi}$, i.e.,

$$\tilde{\Phi}(p(a, b, \cdots, c; f_1, f_2, \cdots, f_n)) = \tilde{\Phi}(p(a, b, \cdots, c; e_1, e_2, \cdots, e_n)). \tag{14}$$

Let

$$\begin{aligned} & p(a, b, \cdots, c; f_1, f_2, \cdots, f_n) \\ = \;& p_1(a, b, \cdots, c; f_1, f_2, \cdots, f_n) + p_2(a, b, \cdots, c; f_1, f_2, \cdots, f_n) \\ & + \cdots + p_t(a, b, \cdots, c; f_1, f_2, \cdots, f_n) \end{aligned} \tag{15}$$

where each p_i is homogeneous in the letter f_1. Let $g_i = \text{cont}(p_i, f_1)$, suppose

$$g_1 > g_2 > \cdots > g_t.$$

Since

$$\begin{aligned} & \tilde{\Phi}(p(a, b, \cdots, c; f_1, f_2, \cdots, f_n)) \\ = \;& \Phi(p(a, b, \cdots, c;\ [\phi(f_1), \phi(f_2), \cdots, \phi(f_n)]^{-1} f_1, f_2, \cdots, f_n)) \\ = \;& [\phi(f_1), \phi(f_2), \cdots, \phi(f_n)]^{-g_1}\Phi(p_1) + \cdots + [\phi(f_1), \phi(f_2), \cdots, \phi(f_n)]^{-g_t}\Phi(p_t), \end{aligned}$$

the condition (14) becomes

$$\begin{aligned} & \Phi([e_1e_2\cdots e_n]^{g_1}(p_1+[f_1f_2\cdots f_n]^{g_1-g_2}p_2+\cdots+[f_1f_2\cdots f_n]^{g_1-g_t}p_t)) \\ = & \Phi([f_1f_2\cdots f_n]^{g_1}(p_1+[e_1e_2\cdots e_n]^{g_1-g_2}p_2+\cdots+[e_1e_2\cdots e_n]^{g_1-g_t}p_t)) \end{aligned} \quad (16)$$

where in the left side $p_i = p_i(a,b,\cdots,c;f_1,f_2,\cdots,f_n)$, and in the right side $p_i = p_i(a,b,\cdots,c;e_1,e_2,\cdots,e_n)$. Let

$$\begin{aligned} & p'(a,b,\cdots,c;f_1,f_2,\cdots,f_n) \\ = & p_1+[f_1f_2\cdots f_n]^{g_1-g_2}p_2+\cdots+[f_1f_2\cdots f_n]^{g_1-g_t}p_t, \end{aligned}$$

then (16) becomes

$$\begin{aligned} & \Phi([e_1e_2\cdots e_n]^{g_1}p'(a,b,\cdots,c;f_1,f_2,\cdots,f_n) \\ = & \Phi([f_1f_2\cdots f_n]^{g_1}p'(a,b,\cdots,c;e_1,e_2,\cdots,e_n). \end{aligned}$$

for all models ϕ such that

$$[\phi(f_1),\phi(f_2),\cdots,\phi(f_n)]\neq 0 \quad \text{and} \quad [\phi(e_1),\phi(e_2),\cdots,\phi(e_n)]\neq 0. \quad (17)$$

Therefore from the Second Fundamental Theorem, we get

$$\begin{aligned} & [e_1e_2\cdots e_n]^{g_1}p'(a,b,\cdots,c;f_1,f_2,\cdots,f_n) \\ = & [f_1f_2\cdots f_n]^{g_1}p'(a,b,\cdots,c;e_1,e_2,\cdots,e_n). \end{aligned} \quad (18)$$

(The condition (17) does not affect the proof for the Second Fundamental Theorem.)

Suppose there is the following order relation

$$e_1<e_2<\cdots<e_n<a<b<\cdots<c<f_1<f_2<\cdots<f_n.$$

By the straightening formula we have

$$p'(a,b,\cdots,c;f_1,f_2,\cdots,f_n)=\sum_i r_i[D_i]$$

where D_i are standard diagrams consisting of the letters $a, b, \cdots, c, f_1, f_2, \cdots, f_n$. Also, by straightening $p(a, b, \cdots, c; e_1, e_2, \cdots, e_n)$ we get

$$p'(a, b, \cdots, c; e_1, e_2, \cdots, e_n) = \sum_i s_i[E_i]$$

where E_i are standard diagrams consisting of the letters $a, b, \cdots, c, e_1, e_2, \cdots, e_n$. From (18), we therefore obtain

$$[e_1 e_2 \cdots e_n]^{g_1} \sum_i r_i[D_i] = [f_1 f_2 \cdots f_n]^{g_1} \sum_i s_i[E_i]. \tag{19}$$

For each i, let $\tilde{D}_i$ be the diagram obtained from D_i by adding g_1 rows of $e_1 e_2 \cdots e_n$ to the top of D_i, and let $\tilde{E}_i$ be the diagram obtained from E_i by adding g_1 rows of $f_1 f_2 \cdots f_n$ to the bottom of E_i, then (19) becomes

$$\sum_i r_i[\tilde{D}_i] = \sum_i s_i[\tilde{E}_i]. \tag{20}$$

Both sides of (20) are linear combinations of standard tableaux. Since standard tableaux form a linear basis of Bracket$[L]$, we infer that the bottom g_1 rows of $\tilde{D}_i$ are all $f_1 f_2 \cdots f_n$, which is also true for D_i, i.e., for each i,

$$[D_i] = [D_i'][f_1 f_2 \cdots f_n]^{g_1}$$

for some diagram D_i' which contains only the letters $a, b, \cdots, c$.

Therefore

$$\begin{aligned} & p'(a, b, \cdots, c; f_1, f_2, \cdots, f_n) = \sum_i r_i[D_i] \\ = \ & [f_1 f_2 \cdots f_n]^{g_1} q(a, b, \cdots, c) \end{aligned} \tag{21}$$

for some bracket polynomial $q(a, b, \cdots, c)$.

Recall that

$$\begin{aligned} & p'(a, b, \cdots, c; f_1, f_2, \cdots, f_n) \\ = \ & p_1 + [f_1 f_2 \cdots f_n]^{g_1 - g_2} p_2 + \cdots + [f_1 f_2 \cdots f_n]^{g_1 - g_t} p_t, \end{aligned} \tag{22}$$

by (21) both sides of the above identity are homogeneous bracket polynomials in the total degree, which is ng_1, of the letters $f_1, f_2, \cdots, f_n$. We call such a polynomial f-homogeneous (with relative f-degree ng_1).
Since $p(a, b, \cdots, c; f_1, f_2, \cdots, f_n)$ is a polynomial in brackets

$$[f_1 \cdots f_{j-1} a f_{j+1} \cdots f_n], \ [f_1 \cdots f_{j-1} b f_{j+1} \cdots f_n], \ \cdots, \ [f_1 \cdots f_{j-1} c f_{j+1} \cdots f_n],$$

which is homogeneous in the letters $a, b, \cdots, c$, we know that the polynomial $p(a, b, \cdots, c; f_1, f_2, \cdots, f_n)$ is also f-homogeneous with relative f-degree h, for some non-negative integer h. By the definition of p_i in (15), we know that each p_i is also f-homogeneous with relative f-degree h.
Then in (22), each term $[f_1 f_2 \cdots f_n]^{g_1 - g_i} p_i$ is f-homogeneous with relative f-degree $n(g_1 - g_i) + h$ which are different for all i, since $g_1 > g_2 > \cdots > g_t$. Therefore in order that both sides of (22) are f-homogeneous polynomials with relative f-degree ng_1, we must have

$$t = 1 \qquad \text{and} \qquad h = g_1.$$

We conclude that

$$\begin{aligned} & p(a, b, \cdots, c; f_1, f_2, \cdots, f_n) \\ = \ & p_1(a, b, \cdots, c; f_1, f_2, \cdots, f_n) \\ = \ & p'(a, b, \cdots, c; f_1, f_2, \cdots, f_n) \\ = \ & [f_1 f_2 \cdots f_n]^{g_1} q(a, b, \cdots, c) \end{aligned}$$

as desired. ∎

Remark. As a corollary of the proof, we know that the polynomial $p(a, b, \cdots, c; f_1, f_2, \cdots, f_n)$ is an invariant if and only if it satisfies

$$\begin{aligned} & [e_1 e_2 \cdots e_n]^g p(a, b, \cdots, c; f_1, f_2, \cdots, f_n) \\ = \ & [f_1 f_2 \cdots f_n]^g p(a, b, \cdots, c; e_1, e_2, \cdots, e_n). \end{aligned}$$

for any set of negative letters $e_1, e_2, \cdots, e_n$, and for some integer g, see (18).

The proceeding results can be read as follows: let p be a polynomial in the coordinates of vectors $a, b, ..., c$ which takes the same value in every unimodular coordinate system. Then p can be expressed as a bracket polynomial. The first fundamental theorem of invariant theory is usually expressed in this form. Note that such a version of the theorem is, strictly speaking , incorrect. Any correct statement of the theorem requires the introduction of syntactic notions, as we have done by introducing the "abstract" algebra of brackets.

9 Supersymmetric Models.

We shall now extend the concept of a model to the supersymmetric bracket algebra, where the alphabet L now is arbitrary. Despite the fact that we now have letters of two signatures, the models still turn out to be Peano spaces.

We briefly digress to recall the notion of a skew-symmetric tensor in a Peano space. Classically, a skew-symmetric tensor of step k is a homogeneous element of degree k in the exterior algebra $\text{Ext}(V)$ over the vector space V. It is a linear combination of decomposable skew-symmetric tensors:

$$t = \sum_i t_i$$

where each t_i is of the form $v_1^{(i)} v_2^{(i)} ... v_k^{(i)}$ for some (anticommutative) vectors $v_1^{(i)}, v_2^{(i)}, ..., v_k^{(i)}$. It thus makes sense , in a Peano space, to write

$$[t, x_1, x_2, ..., x_{n-k}] = \sum_i [v_1^{(i)}, v_2^{(i)}, ..., v_k^{(i)}, x_1, x_2, ..., x_{n-k}]$$

for any vectors $x_1, x_2, ..., x_{n-k}$. In this way, the structure of a Peano space allows us to associate a multilinear form of order $n-k$ to a skew symmetric tensor of step k. Reasoning along these lines, one sees that expressions like

$$\sum_t [t_{(1)}, t'][t_{(2)}, t'']$$

make sense, where t and t' are skew-symmetric tensors of any step. In other words, one can use the bracket notation for skew-symmetric tensors, and this we shall do. A skew symmetric tensor of step k will be a multilinear form of order $n-k$, written $[t, x_1, x_2, ..., x_{n-k}]$. Such skew-symmetric forms form a Hopf algebra.

We now give the definition of a model of a bracket algebra Bracket$[L]$. First we assign each letter a of L a positive integer k where $1 \leq k \leq n$. This integer k is called the arity of a and denoted as arity$(a) = k$. We further require that arity$(a) = 1$ for all negative letters a of L. A **model** of the algebra Bracket$[L]$ in a Peano space $(V, [\ \ ])$ is a function ϕ assigning a vector $\phi(a)$ to every negatively signed letter a, and assigning a skew symmetric tensor $t = \phi(a)$ of step k to every positive letter a of arity k.

Our objective is now to extend the map ϕ to an induced map Φ from Bracket$[L]$ to K, much as in the classical case.

Denote $\otimes^r$Super$[L]$ as the tensor product Super$[L]$ $\otimes$ Super$[L]$ $\otimes \cdots \otimes$ Super$[L]$ (r-times), and similarly for the notion $\otimes^r$Ext$[L]$.

First for each integer r we extend the map ϕ to an induced linear map

$$\Phi_r : \ \otimes^r \mathrm{Super}[L] \ \rightarrow \ \otimes^r \mathrm{Ext}(V)$$

in the following way.

(1) Suppose arity$(a) = k \leq n$ and $\phi(a) = t + t' + \cdots + t''$ where $t, t', ..., t''$ are decomposable skew symmetric tensors of step k.

If $i_1 + i_2 + \cdots + i_r = k$ we set

$$\begin{aligned}&\Phi_r(a^{(i_1)} \otimes a^{(i_2)} \otimes \cdots \otimes a^{(i_r)}) \\ = \; & \sum_t t_{(1)} \otimes \cdots \otimes t_{(r)} + \sum_{t'} t'_{(1)} \otimes \cdots \otimes t'_{(r)} + \cdots + \sum_{t''} t''_{(1)} \otimes \cdots \otimes t''_{(r)}\end{aligned}$$

where the sums range over the terms such that

$$\text{step}(t_{(j)}) = \text{step}(t'_{(j)}) = \cdots = \text{step}(t''_{(j)}) = i_j;$$

if $i_1 + i_2 + \cdots + i_r \neq k$ we set

$$\Phi_r(a^{(i_1)} \otimes a^{(i_2)} \otimes \cdots \otimes a^{(i_r)}) = 0.$$

(2) Fix a linear ordering among the positive letters L^+. For a monomial

$$m = w_1 \otimes w_2 \otimes \cdots \otimes w_r$$

where $w_i \in \text{Div}(L)$, let

$$\text{disp}(w_1 w_2 \cdots w_r) = x_1 x_2 \cdots x_s$$

which is a word in $\text{Mon}(L)$, where $x_i \in L$. Define the number of inversions $i(m)$ of the monomial m by

$$\begin{aligned} i(m) &= \#\{\, (x_i, x_j) \mid i < j, \;\; x_i > x_j, \;\; x_i, x_j \in L^+ \,\} \\ &+ \#\{\, (x_i, x_j) \mid i < j, \;\; x_i \in L^-, \;\; x_j \in L^+ \,\}. \end{aligned}$$

Then for two monomials $m, m' \in \otimes^r \text{Super}[L]$ with $\text{supp}^+(m) \cap \text{supp}^+(m') = \emptyset$ we set

$$\boxed{(-1)^{i(mm')}\Phi_r(mm') = (-1)^{i(m)}\Phi_r(m) \cdot (-1)^{i(m')}\Phi_r(m').}$$

With each Φ_r defined above, we may define an induced map

$$\Phi : \;\; \text{Bracket}[L] \;\to\; K$$

as follows. Given words $w_1, w_2, ..., w_r$ of length n in $\text{Div}(L)$, suppose

$$\Phi_r(w_1 \otimes w_2 \otimes \cdots \otimes w_r) = \sum_i t_i \otimes t'_i \otimes \cdots \otimes t''_i$$

where $t_i, t'_i, ..., t''_i$ are skew symmetric tensors of step n in Ext(L). We set

$$\boxed{\Phi(\text{vert}[w_1, w_2, ..., w_r]) = \textstyle\sum_i \text{vert}[t_i, t'_i, ..., t''_i]}$$

where we define

$$\text{vert}[t_i, t'_i, ..., t''_i] = (-1)^{\frac{r(r-1)}{2}}[t_i][t'_i]\cdots[t''_i].$$

Proposition 14 *Φ_r is a well defined operator.*

Proof. Let

$$m = w_1 \otimes w_2 \otimes ... \otimes w_r \quad \text{and} \quad m' = w'_1 \otimes w'_1 \otimes ... \otimes w'_r$$

be two monomials in $\otimes^r \text{Super}[L]$ with $\text{supp}^+(m) \cup \text{supp}^+(m') = \emptyset$. We denote

$$l_i = \text{length}(w_i) \quad \text{and} \quad l'_i = \text{length}(w'_i)$$

We know that

$$mm' = (-1)^{(|w_1|+...+|w_r|)(|w'_1|+...+|w'_r|)}m'm. \tag{23}$$

It is sufficient to prove that Φ_r preserves the identity (23). Let

$$\Phi_r(m) = \sum_i t_{1i} \otimes ... \otimes t_{ri} \quad \text{and} \quad \Phi_r(m') = \sum_i t'_{1i} \otimes ... \otimes t'_{ri}$$

where t_{ij} and t'_{ij} are skew symmetric tensors of steps l_i and l'_i.

Then

$$\begin{aligned} (-)^{i(mm')}\Phi_r(mm') &= (-1)^{i(m)}\Phi_r(m) \cdot (-1)^{i(m')}\Phi_r(m') \\ &= (-1)^{i(m)+i(m')}(\sum_i t_{1i} \otimes ... \otimes t_{ri}) \cdot (\sum_i t'_{1i} \otimes ... \otimes t'_{ri}), \end{aligned}$$

and

$$\begin{aligned} &(-)^{i(mm')}\Phi_r(mm') = (-1)^{i(m')}\Phi_r(m') \cdot (-1)^{i(m)}\Phi_r(m) \\ &= (-1)^{i(m)+i(m')}(\sum_i t'_{1i} \otimes ... \otimes t'_{ri}) \cdot (\sum_i t_{1i} \otimes ... \otimes t_{ri}) \\ &= (-1)^{i(m)+i(m')} \cdot (-1)^{(l_1+...+l_r)(l'_1+...+l'_r)}(\sum_i t_{1i} \otimes ... \otimes t_{ri}) \cdot (\sum_i t'_{1i} \otimes ... \otimes t'_{ri}), \end{aligned}$$

therefore

$$(-1)^{i(mm')}\Phi_r(mm') = (-1)^{i(m'm)}(-1)^{(l_1+...+l_r)(l'_1+...+l'_r)}\Phi_r(m'm).$$

So in order to prove that Φ_r preserves the identity (23), it is enough to show that

$$\begin{aligned} & (|w_1| + ... + |w_r|)(|w'_1| + ... + |w'_r|) \\ \equiv\ & i(mm') + i(m'm) + (l_1 + ... + l_r)(l'_1 + ... + l'_r) \pmod 2 \end{aligned} \tag{24}$$

Let

$$\mathrm{disp}(w_1w_2...w_r) = x_1x_2...x_p \quad \text{and} \quad \mathrm{disp}(w'_1w'_2...w'_r) = y_1y_2...y_q$$

where x_i, y_i are letters in L, and $\mathrm{supp}^+(x_1x_2...x_p) \cap \mathrm{supp}^+(y_1y_2...y_q) = \emptyset$. Consider the set

$$A = \{(x_i, y_j) \mid 1 \le i \le p, \quad 1 \le j \le q\}$$

and its subsets

$$\begin{aligned} A_1 &= \left\{(x_i, y_j) \mid x_i, y_j \in L^-\right\}, \\ A_2 &= \left\{(x_i, y_j) \mid x_i, y_j \in L^+, \ \ x_i > y_j\right\}, \\ A_3 &= \left\{(x_i, y_j) \mid x_i, y_j \in L^+, \ \ x_i < y_j\right\}, \\ A_4 &= \left\{(x_i, y_j) \mid x_i \in L^-, \ \ y_j \in L^+\right\}, \\ A_5 &= \left\{(x_i, y_j) \mid x_i \in L^+, \ \ y_j \in L^-\right\}. \end{aligned}$$

Since $\mathrm{supp}^+(x_1x_2\cdots x_p) \cap \mathrm{supp}^+(y_1y_2\cdots y_q) = \emptyset$, we get

$$A = A_1 \cup A_2 \cup A_3 \cup A_4 \cup A_5.$$

Furthermore from the definitions we have

$$\begin{aligned} &|A| = (l_1 + l_2 + \cdots + l_r)(l'_1 + l'_2 + \cdots + l'_r), \\ &|A_1| = (|w_1| + |w_2| + \cdots + |w_r|)(|w'_1| + |w'_2| + \cdots + |w'_r|), \\ &|A_2| + |A_4| = i(mm') - i(m) - i(m'), \\ &|A_3| + |A_5| = i(m'm) - i(m') - i(m). \end{aligned}$$

Therefore identity (24) is proved. ∎

The operators Φ_r can be also described by polarization operators. This we proceed now. For a model ϕ, we identify negative letters $a \in L^-$ with the vectors $\phi(a)$ in V. We may suppose L^- is large enough to contain all vectors in V. For a positive letter a of arity k, let

$$\begin{aligned} \phi(a) &= t + t' + \cdots + t'' \\ &= v_1 v_2 \cdots v_k + v_1' v_2' \cdots v_k' + \cdots + v_1'' v_2'' \cdots v_k'' \end{aligned}$$

where $v_i, v_i', ..., v_i''$ are vectors in V. Then the operator

$$\begin{aligned} \mathbf{T^a} &= T(t,a) + T(t',a) + \cdots + T(t'',a) \\ &= D(v_1,a)D(v_2,a)\cdots D(v_k,a) + \cdots + D(v_1'',a)D(v_2'',a)\cdots D(v_k'',a) \end{aligned}$$

is well defined on Super$[L]$. Now we define an operator

$$\mathbf{T_r}: \quad \otimes^r \text{Super}[L] \to \otimes^r \text{Ext}(V)$$

in the following way.

Let m be a monomial in $\otimes^r$Super$[L]$, and let $a, b, ..., c$ be a list of all positive letters occurring in m in the order $a < b < \cdots < c$, then we set

$$\mathbf{T_r}(m) = \mathbf{T^a} \circ \mathbf{T^b} \circ \cdots \circ \mathbf{T^c}(m).$$

Proposition 15 *With the operators Φ_r and $\mathbf{T_r}$ defined above, we have*

$$\boxed{\Phi_r = \mathbf{T_r}.}$$

Proof. Let a be a positive letter of arity k, and

$$\phi(a) = t + t' + \cdots + t''$$

where $t, t', ..., t''$ are decomposable tensors of step k. Then by definition

$$\begin{aligned} &\mathbf{T_r}(a^{(i_1)} \otimes \cdots \otimes a^{(i_r)}) \\ =\ & T(t,a)(a^{(i_1)} \otimes \cdots \otimes a^{(i_r)}) + \cdots + T(t'',a)(a^{(i_1)} \otimes \cdots \otimes a^{(i_r)}) \\ =\ & \sum_t t_{(1)} \otimes \cdots \otimes t_{(r)} + \cdots + \sum_{t''} t''_{(1)} \otimes \cdots \otimes t''_{(r)} \end{aligned}$$

if $i_1 + i_2 + \cdots i_r = k$, and

$$\mathbf{T_r}(a^{(i_1)} \otimes \cdots \otimes a^{(i_r)}) = 0$$

if $i_1 + i_2 + \cdots i_r \neq k$.

Now let m and m' be two monomials in $\otimes^r \mathrm{Super}[L]$ with

$$\mathrm{supp}^+(m) \cap \mathrm{supp}^+(m') = \emptyset,$$

the only thing left to show is

$$(-1)^{i(mm')}\mathbf{T_r}(mm') = (-1)^{i(m)}\mathbf{T_r}(m) \cdot (-1)^{i(m')}\mathbf{T_r}(m').$$

Suppose

$$m = w_1 \otimes w_2 \otimes \cdots \otimes w_r \quad \text{and} \quad m' = w_1' \otimes w_2' \otimes \cdots \otimes w_r',$$

let

$$l_i = \mathrm{length}(w_i) \quad \text{and} \quad l_i' = \mathrm{length}(w_i').$$

The sets A, A_1, A_2, A_3, A_4 are defined as in the proof of Proposition 14. Keep the rule

$$\mathbf{T^a} \circ \mathbf{T^b} = (-1)^{k+l}\mathbf{T^b} \circ \mathbf{T^a}$$

where $\mathrm{arity}(a) = k$ and $\mathrm{arity}(b) = l$ in mind, we have

$$\begin{aligned}
& (-1)^{i(mm')}\mathbf{T_r}(mm') \\
= \; & (-1)^{i(mm')}(-1)^{|A_2|}\mathbf{T^a} \circ \mathbf{T^b} \circ \cdots \circ \mathbf{T^c} \circ \mathbf{T^{a'}} \circ \mathbf{T^{b'}} \circ \cdots \circ \mathbf{T^{c'}}(mm') \\
= \; & (-1)^{i(mm')+|A_2|}(-1)^{|A_4|}\mathbf{T^a} \circ \mathbf{T^b} \circ \cdots \circ \mathbf{T^c}(m) \cdot \mathbf{T^{a'}} \circ \mathbf{T^{b'}} \circ \cdots \circ \mathbf{T^{c'}}(m') \\
= \; & (-1)^{i(mm')+|A_2|+|A_4|}\mathbf{T_r}(m) \cdot \mathbf{T_r}(m') \\
= \; & (-1)^{i(m)+i(m')}\mathbf{T_r}(m) \cdot \mathbf{T_r}(m')
\end{aligned}$$

which completes the proof. ∎

Now we discuss the well-definedness of the map

$$\Phi : \text{Bracket}[L] \to K.$$

Let $w_1, w_2, ..., w_r$ be monomials of length n in Super$[L]$, let a and b be any two letters in L. Comparing the rules

$$D^{(k)}(b,a)\begin{bmatrix} w_1 \\ w_2 \\ \vdots \\ w_r \end{bmatrix} = \sum_{i_1+\cdots+i_r=k} \varepsilon(i_1, i_2, ..., i_r)\begin{bmatrix} D^{(i_1)}(b,a)w_1 \\ D^{(i_2)}(b,a)w_2 \\ \vdots \\ D^{(i_r)}(b,a)w_r \end{bmatrix}$$

and

$$\begin{aligned} &D^{(k)}(b,a)w_1 \otimes w_2 \otimes \cdots \otimes w_r \\ = &\sum_{i_1+\cdots+i_r=k} \varepsilon(i_1, i_2, ..., i_r) D^{(i_1)}(b,a)w_1 \otimes D^{(i_2)}(b,a)w_2 \otimes \cdots \otimes D^{(i_r)}(b,a)w_r, \end{aligned}$$

where

$$\varepsilon(i_1, i_2, ..., i_r) = (-1)^{(|a|+|b|)(i_2|w_1|+i_3(|w_1|+|w_2|)+\cdots+i_r(|w_1|+\cdots+|w_{r-1}|))},$$

We can see that the map

$$\Phi : \text{Bracket}[L] \to K$$

is nothing but a composition of polarizations by Proposition 15. Therefore the well-definedness of Φ follows.

The proceeding definition is easier than it has been made to sound. It can be summarized as follows. To find a image of a letter a of arity k, which appears in different bracket factors of a monomial, one polarizes the "tensor" $a^{(k)}$ to a sum of decomposable skew-symmetric tensors, and then one uses the definition of Φ for negatively signed letters.

Example 1. If $a, b, ..., c$ are positive letters of arity one, and if $a < b < ... < c$ then

$$\Phi([ab...c]) = \Phi([ba...c]) = [\phi(a)\phi(b)...\phi(c)].$$

Example 2. Suppose all letters in $[w][w']\cdots[w'']$ are positively signed and have arity 1, where $w, w', ..., w''$ are words in $\mathrm{Mon}(L)$. Suppose further that the letters of the word $ww'...w''$ occur in increasing order. Then

$$\Phi([w][w']\cdots[w'']) = \Phi([w])\Phi([w'])...\Phi([w''])$$

if n is even, and

$$\Phi([w][w']\cdots[w'']) = (-1)^{\frac{k(k-1)}{2}}\Phi([w])\Phi([w'])...\Phi([w''])$$

if n is odd, where k is the number of brackets.

Example 3. Let $a, b, ..., c$ be k positive letters in the order $a < b < ... < c$, then

$$\Phi([a^{(n)}])[b^{(n)}]...[c^{(n)}]) = \Phi([a^{(n)}]\Phi([b^{(n)}])\cdots\Phi([c^{(n)}])$$

if n is even, and

$$\Phi([a^{(n)}][b^{(n)}]...[c^{(n)}]) = (-1)^{\frac{k(k-1)}{2}}\Phi([a^{(n)}])\Phi([b^{(n)}])\cdots\Phi([c^{(n)}])$$

if n is odd.

Example 4. If $\phi(a) = t, \phi(b) = t'$ and $\phi(c) = t''$, each letter having arity 2, then

$$\Phi([a^{(2)}b][bc^{(2)}]) = -\sum_{t'}[t, t'_{(1)}][t'_{(2)}, t'']$$

and

$$\Phi([a^{(2)}c][cb^{(2)}]) = -\sum_{t'}[t, t''_{(1)}][t''_{(2)}, t'],$$

regardless of the order of a, b, c.

Example 5. With the proper arities and the alphabetical order, we have

$$\Phi([a^{(2)}d][bc^{(2)}]) = [t, y][x, t'']$$

and

$$\Phi([a^{(2)}b][dc^{(2)}]) = -[t, x][y, t'']$$

where $\phi(b) = x$ and $\phi(d) = y$, and where all the letters are positively signed.

10 Fundamental Theorems in Invariant Theory

Our objective is to prove that every statement about skew-symmetric tensors which is independent of a choice of a coordinate system, that is every *geometric* statement, can be expressed as the vanishing property of bracket polynomials. This program requires, first, an extension of the second fundamental theorem of invariant theory to the supersymmetric case; second, an extension of the first fundamental theorem to the supersymmetric case, stating that all occurrences of coordinates can be "eliminated" in any invariant statement. More precisely, we have:

Theorem 2 *(Second fundamental theorem of invariant theory, a version for supersymmetric bracket algebra.) Let p be a polynomial in Bracket[L] such that $\Phi(p) = 0$ for every skew-symmetric model ϕ. Then $p = 0$.*

Proof. We use the same polarization technique as in proving the uniqueness part of Theorem 1 (the Standard Basis Theorem).
Suppose $p \neq 0$, then it can be written uniquely as a sum of standard tableaux

$[D_i]$ with non-zero coefficients,

$$p = \sum_i c_i[D_i]. \tag{25}$$

We may assume that the tableaux $[D_i]$ have the same content. For each positive letter a of arity k occurring in D_i, we can find a suitable set of negative letters $b_1, b_2, ..., b_k$ such that when we polarize away all the positive letters in (25) the resulting polynomial p' is still a sum of standard tableaux with non-zero coefficients. Since p' contains only negative letters, from the classical version of the second fundamental theorem, we know that there exists a model ϕ' such that $\Phi'(p') \neq 0$.

Let ϕ be a model such that

(1) $\phi(a) = \phi'(a)$ if $a \in L^-$,

(2) $\phi(a) = \phi'(b_1)\phi'(b_2)\cdots\phi'(b_k)$, if a is a positive letter of arity k occurring in p, and where $b_1, b_2, ..., b_k$ is the set of negative letters used for the letter a in the polarization process mentioned above.

Then by Proposition 15, we have

$$\Phi(p) = \pm\Phi'(p') \neq 0,$$

the proof is complete. ∎

To state the second result, we require the notion of an **invariant** of a set of letters $a, b, ..., c$ (positive or negative) of arities $i, j, ..., k$ (with the understanding that negative letters are assigned arity 1).

If a is a letter of arity i, we define the **Plüker coordinates** of the letter a relative to the negative letters $f_1, f_2, ..., f_n$ to be

$$\boxed{[a^{(i)} f_1 f_2 \dots f_{h_1 - 1} f_{h_1 + 1} \dots f_{h_i - 1} f_{h_i + 1} \dots f_n]}$$

running through all i-tuples $h_1, h_2, ..., h_i$ of the letters $f_1, f_2, ..., f_n$. Notice that the Plüker coordinates are elements of Bracket$[L]$, not elements of the field K.

An **invariant** is a bracket polynomial $p(a, b, ..., c; f_1, f_2, ..., f_n)$ in the Plüker coordinates of the letters $a, b, ..., c$, homogeneous in total degree of the letters $a, b, ..., c$, such that for all unimodular models ϕ and ϕ' relative to $f_1, f_2, \cdots, f_n$ such that

$$\phi(a) = \phi'(a), \quad \phi(b) = \phi'(b), \quad ..., \quad \phi(c) = \phi'(c)$$

we have

$$\Phi(p) = \Phi'(p).$$

We now state

Theorem 3 *(First fundamental theorem of invariant theory, a version for supersymmetric bracket algebra.) Let p be an invariant of the letters $a, b, ..., c$ of arities $i, j, ..., k$. Then there exists a bracket polynomial $q(a, b, ..., c)$ such that*

$$\boxed{p(a, b, ..., c; f_1, f_2, ..., f_n) = [f_1 f_2 ... f_n]^g q(a, b, ..., c)}$$

for some non-negative integer g.

The proof of this theorem is exactly the same as for the classical version of the first fundamental theorem.

References

1. Barnabei, M., Brini, A. & Rota, G.-C., *On the exterior calculus of invariant theory*, J. of Algebra **96** (1985), 120-160.

2. Brini, A., Palareti, A. & Teolis, A., *Gordan-Capelli series in superalgebras*, Proc. Natl. Acad. Sci. USA, **85** (1988), 1330-1333.

3. Brini, A. & Teolis, A., *Young-Capelli symmetrizers in superalgebras*, Proc. Natl. Acad. Sci. USA, **86** (1989), 775-778.

4. Désarménien, J., Kung, J. P. S. & Rota, G.-C., *Invariant Theory, Young bitableaux and combinatorics,* Adv. in Math. **27** (1978), 63-92.

5. Dubilet, P., Rota, G.-C. & Stein, J. A., *On the foundations of combinatorial theory IX: Combinatorial method in invariant theory,* Stud. Appl. Math **53** (1976), 185-216.

6. Gelfand, M. & Ponomarev, V. A., *Problems of linear algebra and classification of quadruples of subspaces in a finite-dimensional vector space,* Colloquia Mathematica Societatis Janos Bolyai, (1970) 163-237.

7. Grace, J. H. & Young A., *The algebra of invariants*, Chelsea, New York, 1903.

8. Grassmann, H., *Gesammelte Werke* (6 volls.), Teubner, Leipzig, 1911.

9. Grosshans, F., Rota, G.-C. & Stein, J. A., *Invariant Theory and Superalgebras,* Amer. Math. Soc., Providence, RI **69** (1987).

10. Huang, R. Q., Rota, G.-C. & Stein, J. A., *Supersymmetric algebra, supersymmetric space, and invariant theory*, Annali Scuola Normale Superiore (Volume dedicated to L. Radicati), (1989) Pisa.

11. Koh, S. S. (ed), *Invariant Theory*, Lecture Notes in Math. No. 1278, Springer-Verlag, New York, 1987.

12. Kung, J. P. S. (ed.), *Yung tableaux in combinatorics, invariant theory and algebra,* Academic Press, New York, 1982.

13. Kung, J. P. S. & Rota G.-C., *The invariant theory of binary forms,* Bull. Amer. Math. Soc. **10** (1984), 27-85.

14. Littlewood, D. E., *Invariant theory, tensors and group characters*, Phil. Trans. Royal Soc. A, **239** (1944), 305-365.

15. McMillan, T. & White, N. L., *The dotted straightening algorithm*, to appear in J. of Symb. Comp.

16. McMillan, T. & White, N. L., *Cayley factorization*, in Proc. International Symposium on Symbolic and Algebraic Computations, Rome, 1988.

17. Popov, V., *Constructive invariant theory,* Asterisque **87-88** 303-334.

18. Rota, G.-C. & Stein, J. A., *Standard basis in supersymplectic algebras,* Proc. Natl. Acad. Sci. USA, **86** (1989), 2521-2524.

19. Rota, G.-C. & Stein, J. A., *Supersymmetric Hilbert space,* Proc. Natl. Acad. Sci. USA, **87** (1990), 653-657.

20. Rota, G.-C. & Stein, J. A., *Applications of Cayley algebras,* Colloquio Internazionale sulle Teorie Comninatoriee, Tomo II, Accad. Naz. Lincei, Rome, (1976) 71-97.

21. Rota, G.-C. & Stein, J. A., *Symbolic method in invariant theory,* Proc. Natl. Acad. Sci. USA, **83** (1986), 844-847.

22. Springer, T. A., *Invariant Theory*, Lecture Notes in Math. No. 585, Springer-Verlag, New York, 1977.

23. Sturmfels, B. & White, N. L., *Gröbner bases and invariant theory,* Adv. in Math. **76** (1989), 245-259.

24. Turnbull, H. W., *The projective invariants of four medials*, Proc. Edinburgh Math. Soc., **7** (2) (1942), 55-72.

25. Young, A., *The collected papers of Alfred Young*, University of Toronto-Press, Toronto, 1977.

26. Wallace, A. H., *Invariant matrices and Gorden-Capelli series*, Proc. London Math. Soc. **2** (1952), 98-127.

27. Weitzenböck, *Komplex-Symbolik*, Teubner, Leipzig, 1908.

28. Weitzenböck, *Invariantentheorie*, Noordhoff, Groningen, 1923.

29. White, N. L., *Multilinear Cayley factorization*, to appear in J. of Symb. Comp.

30. Whitely, W., *Logic and invariant theory, I. Invariant theory of projective properties,* Trans. Amer. Math. Soc. **177** (1973), 121-139.

31. Weyl, H. , *The classical groups*, Princeton Univ. Press, Princeton NJ, 1946.

(Received: October 1990)

Acta Applicandae Mathematicae **21**: 247–261, 1990.

Aspects of characteristic-free representation theory of GL_n, and some applications to intertwining numbers.

by David A. Buchsbaum

Department of Mathematics, Brandeis University, Waltham, MA 02154, U.S.A.

§1. *Introduction*

Characteristic-free representation theory has been studied by many people over the past fifteen years in many different ways for many different reasons. In the mid-seventies, Carter and Lusztig [6] defined the Weyl modules for the general linear group over any field, independent of characteristic. Towber, a little later [9], defined the Schur and Weyl modules for GL_n over an arbitrary commutative ring. Using standard monomial theory, Lakshmibai, Musili and Seshadri [8] defined the Schur modules for the classical groups over any commutative ring.

For the kinds of problems that grew out of the study of resolutions of determinantal ideals, it was necessary not only to reproduce the Schur and Weyl modules over an arbitrary commutative ring, but to have a larger category of representations to hand in order to use exact sequences, spectral sequences and other homological tools. In large part, the use of homological methods compensates for the inapplicability to a characteristic-free setting of the usual "classical" combinatorial methods. To illustrate briefly: the Schur and Weyl modules are in general not irreducible, and classical decomposition theorems in terms of these modules are not valid. However, in many cases natural (or equivariant) filtrations of a representation may exist whose associated graded modules are described in terms of Schur and Weyl modules as in the classical theory. As a result of the need for this larger category,

AMS subject classifications (1980). 20G05, 18Gxx
Key words. Schur algebras, intertwining numbers.

the notion of the Schur and Weyl module associated to an arbitrary shape matrix was defined in [5, 1]. In fact, the more general notion of Schur *complex* of a map associated to a shape matrix was defined there; the Weyl and Schur modules result as special cases of the Schur complex. Having this larger class of modules and complexes, we are able to form exact sequences, and apply some standard homological algebra. The remainder of this paper will deal with certain facts about these complexes and modules, with certain problems that have arisen in relation to them, and with an application to the computation of some intertwining numbers. We have tried to choose topics that indicate a strong bridge between representation theory and computational algebra.

§2. *Definitions and basic facts about Schur complexes.*

In this section we quickly recall some basic definitions and facts. The reader is referred to [5] for details.

We let R be a commutative ring, and $\phi: G \to F$ a map between free R-modules of finite type. By $\mathbf{\Lambda}(\phi)$ we mean the Hopf algebra $DG \dot{\otimes} \Lambda F$ (where the dot over $\otimes$ indicates that the degrees of elements of DG affect the sign in multiplication), and by $\mathbf{S}(\phi)$ the Hopf algebra $\Lambda G \otimes SF$. (As usual, S and Λ denote the symmetric and exterior powers. D standards for divided powers.) $\mathbf{\Lambda}(\phi)$ and $\mathbf{S}(\phi)$ are bigraded and have differential maps:

a) $D_k G \otimes \Lambda^\ell F \longrightarrow D_{k-1} G \otimes \Lambda^{\ell+1} F$

b) $\Lambda^k G \otimes S_\ell F \longrightarrow \Lambda^{k-1} G \otimes S_{\ell+1} F$.

The map a) is the composition

$$D_k G \otimes \Lambda^\ell F \to D_{k-1} G \otimes G \otimes \Lambda^\ell F \to D_{k-1} G \otimes F \otimes \Lambda^\ell F \to D_{k-1} G \otimes \Lambda^{\ell+1} F$$

where the first map diagonalizes D_kG, the second maps G to F via ϕ, and the third multiplies F with $\Lambda^{\ell}F$ in ΛF. The map b) is defined similarly. If we denote by $\mathbf{\Lambda}^p(\phi) = \sum_{k+\ell=p} D_kG \otimes \Lambda^{\ell}F$ and $\mathbf{S}_p(\phi) = \sum_{k+\ell=p} \Lambda^kG \otimes S_{\ell}F$, these are the summands of total degree p in $\mathbf{\Lambda}(\phi)$ and $\mathbf{S}(\phi)$, and are subcomplexes with respect to the maps a) and b). The diagonal map $\mathbf{\Lambda}(\phi) \to \mathbf{\Lambda}(\phi) \otimes \mathbf{\Lambda}(\phi)$ and the multiplication map $\mathbf{S}(\phi) \otimes \mathbf{S}(\phi) \to \mathbf{S}(\phi)$ are maps of complexes. Note that $\mathbf{\Lambda}^0(\phi) = 0 \to 0 \to R \to 0 = \mathbf{S}_0(\phi)$ and $\mathbf{\Lambda}^1(\phi) = 0 \to G \to F \to 0 = \mathbf{S}_1(\phi)$.

Now let $A = (a_{ij})$ be an $s \times t$ matrix of integers. We say that A is a *shape matrix* if $a_{ij} = 0$ or 1 for all i,j. For the shape matrix A, we define the sequence $(\mathbf{a}_1,\ldots,\mathbf{a}_s)$ by setting $\mathbf{a}_i = \sum_j a_{ij}$, and the sequence $(\mathbf{b}_1,\ldots,\mathbf{b}_t)$ by setting $\mathbf{b}_j = \sum_i a_{ij}$. Given the map $\phi\colon G \to F$ as above, we define the map

$$d_A(\phi)\colon \Lambda^{\mathbf{a}_1}(\phi) \otimes\cdots\otimes \Lambda^{\mathbf{a}_s}(\phi) \longrightarrow S_{\mathbf{b}_1}(\phi) \otimes\cdots\otimes S_{\mathbf{b}_t}(\phi)$$

as the composite

$$\begin{aligned}
&\mathbf{\Lambda}^{\mathbf{a}_1}(\phi) \otimes\cdots\otimes \mathbf{\Lambda}^{\mathbf{a}_s}(\phi) \to \mathbf{\Lambda}^{a_{11}}(\phi) \otimes\cdots\otimes \mathbf{\Lambda}^{a_{1t}}(\phi) \otimes \mathbf{\Lambda}^{a_{21}}(\phi) \otimes\cdots\otimes \mathbf{\Lambda}^{a_{st}}(\phi) \to \\
&\quad \mathbf{\Lambda}^{a_{11}}(\phi) \otimes \mathbf{\Lambda}^{a_{21}}(\phi) \otimes\cdots\otimes \mathbf{\Lambda}^{a_{s1}}(\phi) \otimes \mathbf{\Lambda}^{a_{12}}(\phi) \otimes\cdots\otimes \mathbf{\Lambda}^{a_{st}}(\phi) \to \\
&\quad S_{a_{11}}(\phi) \otimes S_{a_{21}}(\phi) \otimes\cdots\otimes S_{a_{s1}}(\phi) \otimes S_{a_{12}}(\phi) \otimes\cdots\otimes S_{a_{st}}(\phi) \longrightarrow \\
&\qquad S_{b_1}(\phi) \otimes S_{b_2}(\phi) \otimes\cdots\otimes S_{b_i}(\phi),
\end{aligned}$$

where the first map is the indicated diagonalization, the second a reshuffling, the third the appropriate identifications, and the fourth is the apposite multiplication.

The image of the map $d_A(\phi)$ is the *Schur complex of the map* ϕ *associated to the shape matrix* A, and is denoted by $\mathbf{L}_A(\phi)$. When A is the skew-shape of a skew-partition λ/μ, we write $\mathbf{L}_{\lambda/\mu}(\phi)$. (When μ is the zero partition, λ/μ is just the partition λ. In that case we write $\mathbf{L}_\lambda(\phi)$.)

If $G = 0$, it is easy to see that $\mathbf{L}_A(\phi)$ has only one (possibly) non-zero term. This term is called the *Schur module of* F *of shape* A, and is denoted by L_AF. If $F = 0$, then again $\mathbf{L}_A(\phi)$ has only one non-zero term, and this is called the *Weyl module of* G *of shape* A, written K_AG. If A is the shape of a skew-partition λ/μ, we write $L_{\lambda/\mu}F$, $K_{\lambda/\mu}G$. (Recall that A is the shape matrix of a skew-partition λ/μ if $a_{ij} = 1$ for $\mu_i < j \leq \lambda_i$ and 0 otherwise.)

A basic fact that comes into play at almost every turn is the *standard basis theorem* [5]. This states that for any skew-partition λ/μ, there is a natural filtration of the Schur complex $\mathbf{L}_{\lambda/\mu}(\phi)$ whose associated graded is $\sum L_{\lambda/\sigma}F \otimes K_{\sigma/\mu}G$, where σ runs over all partitions between μ and λ. In terms of tableaux, this theorem can be describes as follows. Let $X = \{x_1,\dots,x_m\}$ and $Y = \{y_1,\dots,y_n\}$ be bases of F and G respectively. Then a linear basis for $(\mathbf{L}_{\lambda/\mu}(\phi))_k$ consists of all tableaux of shape λ/μ filled by $X \cup Y$ such that i) in each row and column, the elements of X precede the elements of Y; ii) in each row, the elements of X are strictly increasing, and the elements of Y are weakly increasing; iii) in each column the elements of X are weakly increasing and the elements of Y are strictly increasing; iv) the number of elements in Y is equal to k. The reader will notice that these tableaux resemble the ones described in some of the other talks at this conference.

§3. *Resolutions of Weyl modules.*

In the preceding section we stressed Schur complexes, and defined the Schur and Weyl modules as special cases. This was done partly for efficiency and partly to point out the connections, via the standard basis theorem, with material discussed here by other speakers. For the rest of the paper, we will concentrate on the modules, although some of the results (and questions) can also be stated for these complexes.

There are many ways to motivate looking for "resolutions" of Weyl and Schur modules. In [1] it was pointed out that the study of $\mathbb{Z}$-forms of rational representations was connected with Ext groups, while in [3] the classical Jacobi-Trudi and Giambelli identities provided the impetus for constructing certain types of resolutions. The motivation we will use in this article is that of computing *intertwining numbers*.

If $\lambda = (\lambda_1,\ldots,\lambda_k)$ is a partition, and d is an integer between one and λ_k, we let μ be the partition $(\lambda_1+d,\lambda_2,\ldots,\lambda_k-d)$. Let $\bar{K}_\lambda\bar{F}$ and $\bar{K}_\mu\bar{F}$ be the indicated Weyl modules over $\mathbb{Z}/(p)$ where p is a rational prime, and $\bar{F}$ a finite-dimensional vector space over $\mathbb{Z}/(p)$. The groups $\mathrm{Ext}^i_{GL\bar{F}}(\bar{K}_\lambda\bar{F},\bar{K}_\mu\bar{F})$ are finite-dimensional vector spaces over $\mathbb{Z}/(p)$ whose dimensions are called the intertwining numbers of $\bar{K}_\lambda\bar{F}$ and $\bar{K}_\mu\bar{F}$. These are the numbers we want to compute.

We have put bars over the F and K because any finite-dimensional $\mathbb{Z}/(p)$-space $\bar{F}$ is isomorphic to $\mathbb{Z}/(p)\otimes F$ where F is a free $\mathbb{Z}$-module, and $\bar{K}_\lambda\bar{F}$ is isomorphic to $\mathbb{Z}/(p)\otimes K_\lambda F$ where $K_\lambda F$ is the indicated Weyl module over $\mathbb{Z}$. Thus our $\mathbb{Z}/(p)$-modules are all induced from $\mathbb{Z}$-modules, and the bar is used to signify this fact.

To avoid the restriction of working in characteristic p, we use the exact sequence

$$\text{(UC)} \qquad 0 \longrightarrow \mathbb{Z}(p) \otimes \mathrm{Ext}^{i}_{GLF}(K_\lambda F, K_\mu F) \longrightarrow \mathrm{Ext}^{i}_{GL\bar{F}}(\bar{K}_\lambda \bar{F}, \bar{K}_\mu \bar{F})$$
$$\longrightarrow \mathrm{Tor}_1^{\mathbb{Z}}(\mathbb{Z}/(p), \mathrm{Ext}^{i+1}_{GLF}(K_\lambda F, K_\mu F)) \longrightarrow 0$$

and observe that $\mathrm{Tor}_1^{\mathbb{Z}}(\mathbb{Z}/(p), M)$ = p-torsion of M for any abelian group M. Thus, we will have the intertwining numbers for any prime p if we can compute the integral Ext groups, $\mathrm{Ext}^{i}_{GLF}(K_\mu F, K_\lambda F)$, for every $\ell \leq 0$.

Now it is almost clear what kinds of resolutions we may be looking for: we want a resolution of $K_\lambda F$

$$\mathbf{X}: \cdots \longrightarrow X_2 \longrightarrow X_1 \longrightarrow X_0 \longrightarrow K_\lambda F \longrightarrow 0$$

such that $\mathrm{Ext}^{i}_{GLF}(K_\lambda F, K_\mu F) = H^i(\mathrm{Hom}_{GLF}(\mathbf{X}, K_\mu F))$.

In order to really make sense of this, we want to replace GLF by a suitable ring, we want the X_i to be projective modules over that ring, and we want the groups $\mathrm{Hom}_{GLF}(\mathbf{X}, K_\mu F)$ to be somewhat computable. This involves introducing the *Schur algebra* and *weight modules* as discussed in [7] and [2]. For those not familiar with these notions, we will highlight the main points here. All details can be found in the references cited above.

For any commutative ring R and free R-module F of finite type, the symmetric group Γ_s operates (on the right) on the s-fold tensor product $F \otimes \cdots \otimes F = \otimes^s F$. Define the *Schur algebra of degree* s to be the algebra of Γ_s-endomorphisms contained in $\mathrm{End}_R(\otimes^s F)$. If we let $E = \mathrm{End}_R(F)$, it is easy to see that this Schur algebra, denoted A_s, is (as a module) isomorphic to $D_s E$, where D signifies divided power. More importantly, it is the universal enveloping algebra for all homogeneous polynomial representations of GLF of degree s. Thus, in particular, if the partition λ that we started with is of weight s (i.e. $\sum \lambda_i = s$), then $K_\lambda F$ and $K_\mu F$ are A_s-modules, and we may replace the computation of $\mathrm{Ext}^i_{GLF}(K_\lambda F, K_\mu F)$ by that of $\mathrm{Ext}^i_{A_s}(K_\lambda F, K_\mu F)$. Consequently, the resolution $\mathbf{X}$ that we are looking for should have all of the X_i projective modules over A_s (we are of course considering our ground ring to be $\mathbb{Z}$).

Fortunately, the following theorem turns out to be true. If F has rank n, then the Schur algebra A_s decomposes as a direct sum of left ideals into the sum $\sum D_{a_1}F \otimes \cdots \otimes D_{a_n}F$ where the sum is taken over all sequences $(a_1, \ldots, a_n)$ of non-negative integers such that $\sum a_1 = s$ (i.e. all weights of GLF of degree s). Thus all the modules $D_{a_1}F \otimes \cdots \otimes D_{a_n}F$ are A_s-projective. It also turns out that for any A_s-module M, $\mathrm{Hom}_{A_s}(D_{a_1}F \otimes \cdots \otimes D_{a_n}F, M)$ is the weight submodule of M corresponding to the weight $(a_1, \ldots, a_n)$. Therefore, if M has a basis given by standard tableaux (as is the case when $M = K_\mu F$), then this weight submodule of M is the free abelian group generated by those standard tableaux having content $(1^{a_1}, 2^{a_2}, \ldots, n^{a_n})$.

With these facts at our disposal, it now makes sense to ask if we can always find a finite resolution, $\mathbf{X}$, of $K_\lambda F$ such that each term X_i of $\mathbf{X}$ is the direct sum of projective A_s-modules of the form $D_{a_1}F \otimes \cdots \otimes D_{a_n}F$. The answer to this question is yes (see [2]), but it is not immediately obvious that this is so. Rather than belabor the details here, let us illustrate how the attempt to solve this problem leads to the introduction of the larger class of representations mentioned in §1, and how homological techniques replace the standard counting techniques.

From now on, we will draw shapes to represent shape matrices. E.g. 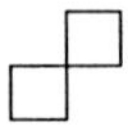 will stand for $\begin{bmatrix} 0 & 1 \\ 1 & 0 \end{bmatrix}$. Also, we will write an exact sequence of shapes to signify either the exact sequence of Schur or Weyl modules corresponding to those shapes. For instance, we will write

(A) 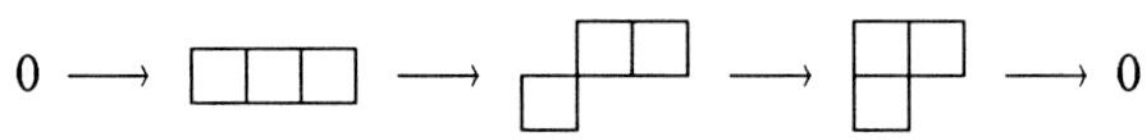

is exact to indicate that the sequences of corresponding Weyl or Schur modules is exact.

Now from the exact sequence (A), we see that $K_{(2,1)}F$ has a resolution:

$$0 \longrightarrow D_3F \longrightarrow D_2F \otimes D_1F \longrightarrow K_{(2,1)}F \longrightarrow 0 .$$

Let us look, however, at $K_{(2,2)}F$. How can we resolve that? Using the same general result that provided the exactness of (A), we get that

(B) 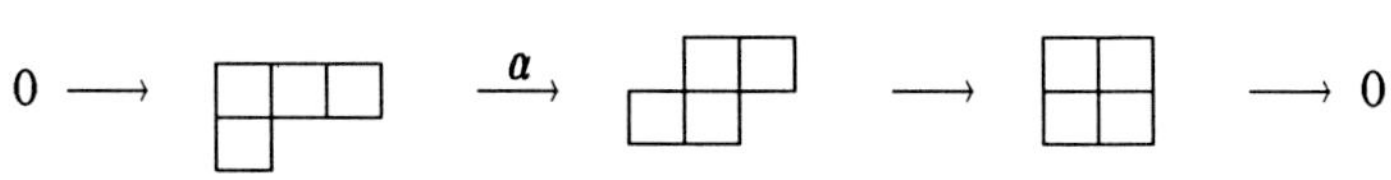

(C) 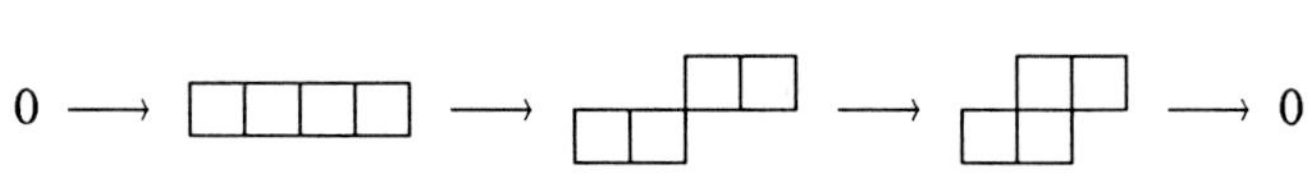

(D) 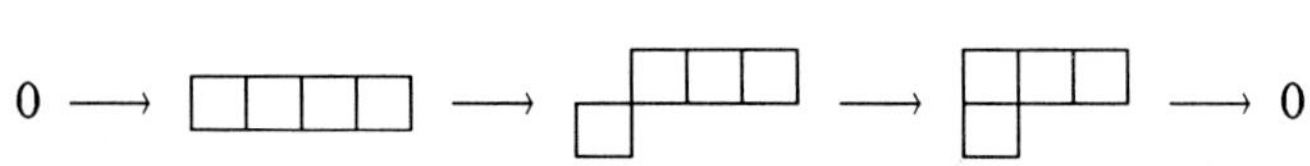

are exact. Thus $K_{(3,2)/(1)}F$ and $K_{(3,1)}F$ have resolutions of the desired type, and because they are projective resolutions, there is a map from the resolution of $K_{(3,1)}F$ into that of $K_{(3,2)/(1)}F$ covering the map α. The mapping cone of this map then provides a resolution of $K_{(2,2)}F$ of the desired type.

Notice that although we have in both cases (A) and (B) been looking for resolutions of *partitions*, we had to consider a resolution of a skew-shape in case (C). Thus, to use this exact sequence - mapping cone approach, even for two-rowed shapes, we have to consider skew-shapes in our category of representations. In fact, what one proves (by induction on the number of overlaps between the first and second rows) is that $K_{\lambda/\mu}F$ has a resolution of the desired type for every skew-partition λ/μ.

Now let us go to three-rowed shapes. Clearly, we must at least consider the category of skew-shapes and not just restrict ourselves to partitions. If we take $\lambda/\mu = (4,3,3)/(1,1)$, we obtain an exact sequence

(E) 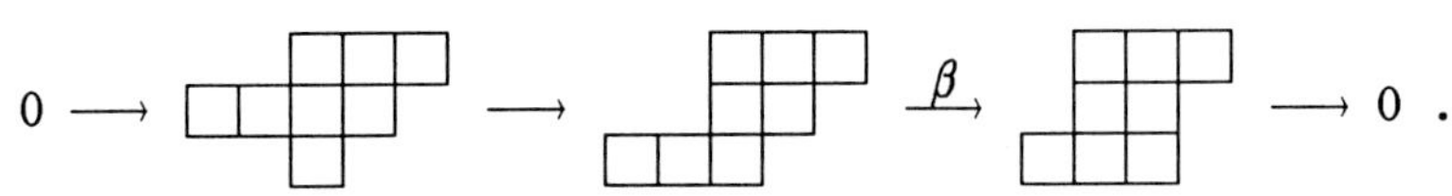

This last shape on the left is not a skew-shape, but unfortunately it *is* the kernel of the map β. Thus, if one still wants to do an induction on total number of overlaps using the kind of exact sequences we have been considering, one must enlarge the category of shapes that we work with.

Of course, as the number of rows increases, the types of different shapes that can appear become more varied. They can, however, all be described as "modified" skew-partitions in the sense that they correspond to pairs of sequences $\mathbf{a} = (a_1,\ldots,a_k)$, $b = (b_1,\ldots,b_k)$ with

$0 \le b_i \le a_i$ for all 1, **a** a partition, $(b_1,\ldots,b_{k-1})$ a partition and such that for some t, $b_t < b_k \le b_{t-1}$ and $a_{k-1} - a_k \ge k-t$. With this enlarged category of shapes, one can prove the exactness of sequences of type (E) and, using the projectivity of the resolutions together with mapping cone constructions, one arrives at the existence of our sought-for resolutions. A similar story can be told for Schur modules and their resolutions in terms of tensor products of exterior powers, but we will not go into that here.

§4. *Some problems concerning Schur and Weyl modules.*

In this section, we will list a few open questions about Schur and Weyl modules. They all apply equally to Schur complexes.

Question 1. It is clear that if A is a shape matrix and A' is obtained from A by adjoining rows or columns of zeros, then the Schur and Weyl modules associated to A and A' are isomorphic. also, if the rows or columns of A are permuted, the corresponding modules are the same. The question then is: Is there a normal form for a shape matrix?

Question 2. In [5] we prove that the Schur and Weyl modules associated to skew-shapes are universally free. It is also easy to prove that the extended class of shapes described in §3 gives us universally free modules. Is the same true for any shape matrix? If not, for what shapes is it true?

Question 3. The standard basis theorem was proved for skew-partitions. At this point, there is not a standard basis theorem for our extended class of shapes. Can one find a general standard basis theorem?

Question 4. Is there a Littlewood-Richardson type of theorem for our extended class of shapes? For any class of shapes besides skew-shapes?

§5. ***Calculation of some intertwining numbers.***

Just for the purpose of illustration, we will consider a two-rowed partition $\lambda = (\lambda_1,\lambda_2)$, a positive integer $d \leq \lambda_2$, the partition $\mu = (\lambda_1+d,\lambda_2-d)$, and calculate the abelian group $\mathrm{Ext}^1_{A_s}(K_\lambda F,K_\mu F)$, where $s = \lambda_1+\lambda_2$. Since $\mathbb{Q} \otimes K_\lambda F$ and $\mathbb{Q} \otimes K_\mu F$ are irreducible rational representations of GLF, it follows that $\mathrm{Hom}_{A_s}(K_\lambda F,K_\mu F) = 0$. Therefore, by the exact sequence (UC) of §3, it follows that for every prime p, the dimension of the $\mathbb{Z}/(p)$-vector space $\mathrm{Hom}_{GL\bar{F}}(\bar{K}_\lambda \bar{F},\bar{K}_\mu \bar{F})$ is equal to the dimension of the p-torsion subgroup of $\mathrm{Ext}^i_{A_s}(K_\lambda F,K_\mu F)$.

We know that we have a resolution

$$\mathbf{X}: \cdots \longrightarrow X_2 \longrightarrow X_1 \longrightarrow X_0 \longrightarrow K_\lambda F \longrightarrow 0$$

where each X_i is the direct sum of tensor products of divided powers of F, and we want to calculate the cohomology of

$$\mathrm{Hom}_{A_s}(X_0,K_\mu F) \xrightarrow{a} \mathrm{Hom}_{A_s}(X_1,K_\mu F) \xrightarrow{b} \mathrm{Hom}_{A_s}(X_2,K_\mu F) .$$

All of these groups are free abelian, so that the kernel of b is a direct summand of $\mathrm{Hom}_{A_s}(X_1,K_\mu F)$. Since $\mathbb{Q} \otimes \mathrm{Ext}^1_{A_s}(K_\lambda F,K_\mu F) = 0$, it follows that $\mathrm{Ext}^1_{A_s}(K_\lambda F, K_\mu F) \approx \mathbb{Z}(e_1) \oplus \cdots \oplus \mathbb{Z}/(e_\ell)$ where $e_1, \ldots, e_\ell$ are the invariant factors of the map a. Thus, we simply have to find the matrix for a, and calculate its invariant factors.

From [5], we know that $X_0 = D_{\lambda_1}F \otimes D_{\lambda_2}F$, and $X_1 = \sum_{k \geq 1} D_{\lambda_1+k}F \otimes D_{\lambda_2-k}F$. The map □: $X_1 \to X_0$ is defined on each $D_{\lambda_1+k}F \otimes D_{\lambda_2-k}F$ to be the composite of

$$D_{\lambda_1+k}F \otimes D_{\lambda_2-k}F \xrightarrow{\Delta \otimes 1} D_{\lambda_1}F \otimes D_kF \otimes D_{\lambda_2-k}F \xrightarrow{1 \otimes m} D_{\lambda_1}F \otimes D_{\lambda_2}F$$

where Δ is the indicated diagonal, and m the indicated multiplication.

To compute a basis for $\mathrm{Hom}_{A_s}(X_0, K_\mu F)$, we have to find all standard tableaux of shape μ having content $(1^{\lambda_1}, 2^{\lambda_2})$. (Standard tableaux for Weyl modules have weakly increasing rows and strictly increasing columns.) It is clear that there is only one such tableau:

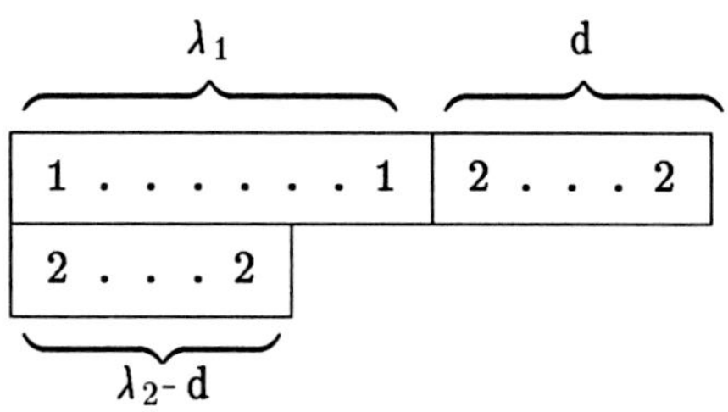

This corresponds to the map a: $D_{\lambda_1}F \otimes D_{\lambda_2}F \to K_\mu F$ defined by $a(X \otimes Y) = \sum \overline{xy(d) \otimes y'(\lambda_2 - d)}$ where $\Delta(y) = \sum y(d) \otimes y'(\lambda_2 - d)$ in $D_dF \otimes D_{\lambda_2-d}F$ is the diagonalization of y, and the bar indicates the image of the barred elements in the Weyl module.

For the computation of $\mathrm{Hom}_{A_s}(X_1, K_\mu F)$, we first note that $\mathrm{Hom}_{A_s}(X_1, K_\mu F) = \sum \mathrm{Hom}_{A_s}(D_{\lambda_1+k}F \otimes D_{\lambda_2-k}F, K_\mu F)$ and for $k > d$, the summands are zero. For $1 \le k \le d$, we get exactly one standard tableau, so $\mathrm{Hom}_{A_s}(X_1, K_\mu F)$ is a free abelian group of rank d with the k-th basis element corresponding to the tableau

$$\beta_k = \begin{array}{|c|c|c|} \hline \overbrace{1 \;.\;.\;.\;.\;.\;.\; 1}^{\lambda_1} & \overbrace{1\;.\;.\;.\;1}^{k} & \overbrace{2\;.\;.\;.\;2}^{d-k} \\ \hline \underbrace{2\;.\;.\;.\;2}_{\lambda_2-d} & & \\ \hline \end{array}$$

To compute the matrix of a, we calculate $a(\alpha) = \alpha \circ \square$ on the elements $1^{(\lambda_1+k)} \otimes 2^{(\lambda_2-k)}$ for $k = 1,\ldots,d$.

$$(*) \qquad \alpha\square(1^{(\lambda_1+k)} \otimes 2^{(\lambda_2-k)}) = \alpha(1^{(\lambda_1)} \otimes 1^{(k)}2^{(\lambda_2-k)})$$

$$= \sum_{\ell=0}^{k} \begin{bmatrix}\lambda_1+\ell \\ \ell\end{bmatrix} \overline{1^{(\lambda_1+\ell)}2^{(d-\ell)} \otimes 1^{(k-\ell)}2^{(\lambda_2-k-d+\ell)}}$$

Note that for $\ell = k$, the over-scored term represents a standard tableau, but for $\ell < k$ we must straighten. To straighten, notice that the box map for $K_\mu F$ (which defines the straightening rule for the Weyl modules), mapping $\sum D_{\lambda_1+d+t}F \otimes D_{\lambda_2-d-t}F \to D_{\lambda_1+d}F \otimes D_{\lambda_2-d}F$ takes the element

$$1^{(\lambda_1+\ell)}2^{(d-\ell)} \otimes 2^{(\lambda_2-k-d+\ell)} \qquad \text{to}$$

$$\sum_{r\ge 0} \begin{bmatrix}\lambda_1-d-r \\ k-\ell-r\end{bmatrix} 1^{(\lambda_1+k-r)}2^{(d-k+r)} \otimes 1^{(r)}2^{(\lambda_2-d-r)} .$$

By induction, it is easy to show that each summand in (*) is equal to

$$(-1)^{k-\ell}\begin{bmatrix}\lambda_2-d\\ k-\ell\end{bmatrix} 1^{(\lambda_1+k)}2^{(d-k)} \otimes 2^{(\lambda_2-d)} = (-1)^{k-\ell}\begin{bmatrix}\lambda_2-d\\ k-\ell\end{bmatrix}\beta_k .$$

Thus the sum in (*) reduces to

$$\left\{\sum (-1)^{k-\ell}\begin{bmatrix}\lambda_1+\ell\\ \ell\end{bmatrix}\begin{bmatrix}\lambda_2-d\\ k-\ell\end{bmatrix}\right\}\beta_k .$$

Since $\begin{bmatrix}\lambda_1+\ell\\ \ell\end{bmatrix} = (-1)^{\ell}\begin{bmatrix}-\lambda_1-\ell\\ \ell\end{bmatrix}$, the coefficient of β_k is $(-1)^k\begin{bmatrix}\lambda_2-\lambda_1-d-1\\ k\end{bmatrix}$ which in turn is $\begin{bmatrix}\lambda_1-\lambda_2+d+1\\ k\end{bmatrix}$. We therefore see that the matrix for a is the $1\times d$ matrix: $\begin{bmatrix}v+d+1\\ 1\end{bmatrix}\begin{bmatrix}v+d+2\\ 2\end{bmatrix} \cdot\cdot\cdot\cdot\cdot \begin{bmatrix}v+2d\\ d\end{bmatrix}$, where $v = \lambda_1 - \lambda_2$ and $\mathrm{Ext}^1_{A_s}(K_\lambda F, K_\mu F) = \mathbb{Z}/(e)$ where $e = v+d+1/\gcd\{v+d+1, \mathrm{lcm}(1,\ldots,d)\}$ (see [4]).

In the three-rowed case, similar calculations have yielded the matrix whose invariant factors we have to compute to obtain the Ext^1 groups, but for $d \geq 3$ we still have not found the right expression for these factors. We did make a conjecture about them in [4], but thanks to P. Valabrega and his computer, we now know that that conjecture is wrong. However, it seems to be not far from the mark, and more use of the computer should at least clear up the three-rowed case. As for the higher Ext groups, we clearly need a more explicit description of resolutions of Weyl modules in order to do our calculations. For two-rowed shapes we do have explicit resolutions, so it would be interesting to carry out this program in that case.

BIBLIOGRAPHY

[1] K. Akin and D. A. Buchsbaum, *Characteristic-free representation theory of the general linear group.* Adv. in Math. **58** (1985), 149-200.

[2] ______________________, *Characteristic-free representation theory of the general linear group, II: Homological considerations.* Adv. in Math. **72** (1988), 171-210.

[3] ______________________, *Characteristic-free realizations of the Giambelli and Jacobi-Trudi determinantal identities.* Proc. of K.I.T. Workshop on Algebra and Toplogy, (1987), 1-19.

[4] ______________________, *Representations, resolutions and intertwining numbers.* Commutative Algebra: Proc. of a microprogram held at MSRI June 15-July 2, 1987. (1989), 1-19.

[5] K. Akin, D. A. Buchsbaum, and J. Weyman, *Schur functors and Schur complexes.* Adv. in Math. **44** (1982), 207-278.

[6] R. W. Carter and G. Lusztig, *On the modular representations of the general linear and symmetric groups.* Math. Z. **136** (1974), 193-242.

[7] J. A. Green, *Polynomial Representations of* GL_n. Lectures Notes in Mathematics, No. 830, Springer-Verlag, Berlin, (1980).

[8] V. Lakshmibai, C. Musili, and C. S. Seshadri, *Cohomology of line bundles on* G/B. Ann. Sci. École Norm. Sup. **7** (1974), 89-138.

[9] J. Towber, *Two new functors from modules to algebras.* J. Algebra **47** (1977), 80-109.

(Received: October 1990)